Partial Differential Equations

Mathematics and Its Applications

Richard Bellman

Dept. of Electrical Engineering, University of Southern California, Los Angeles, U.S.A.; Center for Applied Mathematics, The University of Georgia, Athens, Georgia, U.S.A.

and

George Adomian

Center for Applied Mathematics, The University of Georgia, Athens, Georgia, U.S.A.

Partial Differential Equations

New Methods for Their Treatment and Solution

D. Reidel Publishing Company

A MEMBER OF THE KLUWER ACADEMIC PUBLISHERS GROUP

Dordrecht / Boston / Lancaster

Library of Congress Cataloging in Publication Data

CIP

Bellman, Richard Ernest, 1920-
Partial differential equations.

(Mathematics and its applications)
Bibliography: p.
Includes index.
1. Differential equations, Partial. I. Adomian, G. II. Title.
III. Series: Mathematics and its applications
(D. Reidel Publishing Company)
QA374.B33 1984 515.3'53 84-18241
ISBN-13: 978-94-010-8804-6 e-ISBN-13: 978-94-009-5209-6
DOI: 10.1007/978-94-009-5209-6

Published by D. Reidel Publishing Company
P.O. Box 17, 3300 AA Dordrecht, Holland

Sold and distributed in the U.S.A. and Canada
by Kluwer Academic Publishers
190 Old Derby Street, Hingham, MA 02043, U.S.A.

In all other countries, sold and distributed
by Kluwer Academic Publishers Group,
P.O. Box 322, 3300 AH Dordrecht, Holland

To Peter D. Lax

TABLE OF CONTENTS

PREFACE

The purpose of this book is to present some new methods in the treatment of partial differential equations. Some of these methods lead to effective numerical algorithms when combined with the digital computer. Also presented is a useful chapter on Green's functions which generalizes, after an introduction, to new methods of obtaining Green's functions for partial differential operators. Finally some very new material is presented on solving partial differential equations by Adomian's decomposition methodology. This method can yield realistic computable solutions for linear or non-linear cases even for strong nonlinearities, and also for deterministic or stochastic cases - again even if strong stochasticity is involved. Some interesting examples are discussed here and are to be followed by a book dealing with frontier applications in physics and engineering.

In Chapter I, it is shown that a use of positive operators can lead to monotone convergence for various classes of nonlinear partial differential equations.

In Chapter II, the utility of conservation technique is shown. These techniques are suggested by physical principles. In Chapter III, it is shown that dynamic programming applied to variational problems leads to interesting classes of nonlinear partial differential equations. In Chapter IV, this is investigated in greater detail. In Chapter V, we show that the use of a transformation suggested by dynamic programming leads to a new method of successive approximations.

In Chapter VI, we consider the variation of characteristic values and characteristic functions with the domain, restricting our attention to the one-dimensional case. We point out how the method may be applied to the multidimensional case. In Chapter VII, we apply this method to obtain the classic Hadamard variational formula. It is pointed out that the same method can treat multidimensional variational problems.

In Chapter VIII, we consider the discretized form of the two-dimensional potential equation. This equation arises from the minimization of a quadratic form. The minimization is carried out by means of dynamic programming and the dimensionality difficulties are circumvented observing that the minimum itself is a quadratic form. In Chapter IX, we consider the discretized form of the three-dimensional potential equation. Dimensionality difficulties are now avoided by using a more general domain. In Chapter X, we consider the linear heat equation. Our procedure has two parts. First, we use the Laplace transform, obtaining an equation of potential type. Then, we employ a numerical inversion method. In Chapter XI, we consider nonlinear parabolic partial differential equations, obtaining an analogue of the classical Poincaré-Lyapunov theorem for ordinary differential equations. The results required concerning the linear equation are of interest in themselves.

In Chapter XII, we give some examples of the use of differential quadrature. In Chapter XIII, we consider an adaptive grid to treat nonlinear partial differential equations.

In Chapter XIV, we consider the expansion of the solution of a partial differential equation in orthogonal functions. This yields an infinite system of ordinary differential equations. To obtain numerical results from this system, some method of truncation must be employed. Many interesting stability questions arise in this way.

The concluding chapters deal with Green's functions, methods of determining Green's functions for partial differential equations, and new and valuable methods of dealing with nonlinearity and/or randomness in multidimensional problems using the basic decomposition method first discussed by Adomian in Stochastic Systems (Academic Press, 1983); Stochastic Systems II and also Applications of Stochastic Systems Theory to Physics and Engineering will appear shortly. Chapter XV deals with Green's functions; Chapter XVI shows a method for approximate calculation of Green's functions which is very useful in computation; Chapter XVII deals with Green's functions for partial differential equations; Chapter XVIII deals with the Itô equation and a general stochastic model for dynamical systems. The last chapter is completely new and deals with solution of nonlinear partial differential equations by the decomposition method and provides examples as well. Randolph Rach of Raytheon Company has been very helpful here.

A great deal of work has been done in the field of partial differential equations. We have made no attempt to cover it all. Rather, we have restricted our attention to areas where we have worked ourselves. Even here, as will be apparent from the text, there is still a great deal to be done.

RICHARD E. BELLMAN

PREFACE BY SECOND AUTHOR

Before the final galley corrections to this book, Richard E. Bellman passed into history on March 19, 1984 at the age of 63.

To the very end, his mind was filled with exciting ideas, and he planned further contributions. Having had the privilege of knowing him for over 20 years, and of close collaboration in recent years, it is clear that the loss to science, engineering, and mathematics is immeasurable and profound. Richard Bellman will be remembered around the world as one of the foremost mathematical scientists of this century.

Richard Bellman's influence on present and future researchers and generations of students to come will be pervasive and deep. His incredibly productive pioneering work and creative thought in mathematics and its applications to science, engineering, medicine, and economics have made him a leader in his field. His helpfulness and compassion and courage endeared him to our hearts.

G. ADOMIAN

Chapter I

MONOTONE CONVERGENCE AND POSITIVE OPERATORS

1. INTRODUCTION

Consider the equation

$$L(u) = N(u). \tag{1}$$

Here, L is a linear operator and N is nonlinear. There are three standard ways of treating equations of this type. The fixed-point method, the variational approach, and the method of successive approximations. The first two are nonconstructive and the third is constructive.

The purpose of this chapter is to add a fourth method based on monotone operators. This will be a constructive method which depends upon the method of successive approximations. However, instead of the usual approach we will use monotonicity which will be derived from the monotone character of the linear operator.

2. MONOTONE OPERATORS

We say that a linear operator is a monotone if

$$L(u) \geq L(v) \tag{1}$$

implies

$$u \geq v. \tag{2}$$

Sometimes, this inequality is reversed. It will be clear how the argument may be modified.

Let us assume that N is always positive and monotone increasing. Examples are the functions u^2 and e^u. In many cases all we need is the monotone increasing character.

These concepts can be generalized to the case of symmetric matrices with the ordering of symmetric matrices.

The functions u and v usually satisfy the same boundary conditions. It follows by virtue of the linearity that we may write

$$L(w) \geq 0 \quad \text{implies} \quad w \geq 0. \tag{3}$$

where w is 0 on the boundary.

3. MONOTONICITY

Let us consider the following successive approximations

$$L(u) = 0$$
$$L(u_{n+1}) = N(u_n). \tag{1}$$

By virtue of the assumptions concerning L and N, we have monotonicity.

4. CONVERGENCE

In the previous section, we have derived a sequence which is monotone increasing. It will be convergent if we show that the sequence is uniformly bounded. There are two ways we can proceed.

In the first place, we can impose conditions on N which guarantee this boundedness. In the second place, we often are in the situation where a fixed point or a variational argument established the existence of a solution. Often, the uniqueness of a solution is readily established. The same argument as above shows that this solution is an upper bound. Hence, we have convergence. We shall give an example of this below.

It is worthwhile pointing out that in this case we have convergence wherever a solution exists.

In the foregoing section, we showed that we had monotone convergence. Using Dini's theorem, in many cases we can improve this to uniform convergence.

5. DIFFERENTIAL EQUATIONS WITH INITIAL CONDITIONS

Let us consider

$$u'' + 3u' + 2u = u^2,$$
$$u(0) = c_1, \quad u'(0) = c_2. \tag{1}$$

From the Poincaré-Lyapunov theorem, we know that a solution exists throughout the entire t-interval if c_1 and c_2 are sufficiently small. The linear operator is monotone. This follows from general results, but can readily be established by using factorization and the result for the first order case repeatedly. In the first order case this monotonicity is evident because we have an explicit solution.

6. TWO-POINT BOUNDARY CONDITIONS

Consider the equation

$$u'' - u = u^3$$
$$u(0) = c, \quad u(r) = 0. \tag{1}$$

The monotonicity of the linear operator now depends upon the assumptions.

7. NONLINEAR HEAT EQUATION

Consider the equation

$$u_t - u_{xx} = N(u),$$
$$u(x,0) = \tau(x), \quad 0 \le x \le 1 \tag{1}$$
$$u(0,t) = u(1,t) = 0 \quad t > 0.$$

We assume that $N(u)$ is a power series in u lacking constant and first degree terms. There is an analogue of the Poincaré-Lyapunov theorem for this equation. If we assume the $\tau(x)$ is sufficiently small, the solution exists for all positive t.

The linear operator is monotone. This may be established in many ways. One way is to use the difference equation

$$u(x,t+\Delta) = u(x-\Delta,t) + u(x+\Delta,t)/2 + \Delta f(t). \tag{2}$$

Here, we assume that $f(t)$ is non-negative and that t takes only discrete values $0, \Delta, \ldots$ This difference relation makes the non-negativity of u clear. Passing to the limit as $\Delta \to 0$, we obtain the linear operator.

8. THE NONLINEAR POTENTIAL EQUATION

Consider the equation

$$u_{xx} + u_{yy} = N(u). \tag{1}$$

The equation

$$u_{xx} + u_{yy} = e^u, \tag{2}$$

is an important equation.

The linear operator is monotone. This may be established by using a variational expression as described above.

BIBLIOGRAPHY AND COMMENTS

Section 1. We are following the paper:

R. Bellman, 'Monotone Operators and Nonlinear Equations', Journal of Mathematical Analysis and Applications 67 (1979), 158-162.

Section 6. The arguments may be found in:

R. Bellman, 'On the Non-Negativity of Green's Functions', Boll. D'Unione Matematico 12 (1957), 411-413.

R. Bellman, 'On Variation - Diminishing Properties of Green's Functions', Boll. D'Unione Matematico 16 (1961), 164-166.

Chapter II

CONSERVATION

1. INTRODUCTION

In a series of papers, we have applied invariant imbedding to provide new analytic and computational approaches to a variety of processes of mathematical physics. We began a detailed analysis of the ordinary and partial differential equations of invariant imbedding, concentrating upon existence and uniqueness of solution, nonnegativity of solution, and convergence of associated difference algorithms.

In this chapter, we wish to show that for a quite general class of transport processes involving particle-particle interaction as well as the usual particle-medium interaction we can obtain difference approximations which exhibit nonnegativity and boundedness in an immediate fashion. Furthermore, a uniform Lipschitz condition is preserved.

In the second part of the chapter we examine Riccati matrix equations using conservation techniques.

These conservation techniques are suggested by physical principles. It is usually a matter of conserving a quantity such as mass or energy. Conservation techniques are invariance techniques. These were examined systematically by E. Noether, using the theory of invariants.

It is important to note that we have systematically considered the discretized form of the equations. If we do this, we can handle ordinary differential equations. Conservation now yields bounds on the solution.

2. ANALYTIC AND PHYSICAL PRELIMINARIES

When interaction occurs, the expected fluxes in the transport process do not depend linearly upon the incident fluxes. To apply invariant imbedding techniques we must regard the reflection, transmission, and loss functions as functions of both the length of the rod and the incident intensities. We let

$$r_i(x,y) = \text{the expected flux of particles in state } i \text{ reflected from a rod of length } x \text{ when the incident flux at } x \text{ has intensity } y. \tag{1}$$

Here, the incident intensity y is an N-dimensional vector whose jth component y_j is the intensity of incident flux in state j. For convenience we also introduce the column vector $r(x,y)$ with components $r_i(x,y)$ $(i=1,2,\ldots,N)$.

The transmission and loss vector functions $t(x,y)$ and $l(x,y)$ are defined similarly. We let

$$\begin{aligned} t_i(x,y) &= \text{the expected flux of particles in state } i \text{ transmitted through a rod of length } x \text{ when the incident flux at } x \text{ has intensity } y; \\ \ell_i(x,y) &= \text{the expected flux of particles in state } i \text{ absorbed or annihilated within a rod of length } x \text{ when the incident flux at } x \text{ has intensity } y. \end{aligned} \tag{2}$$

To obtain differential equations for these functions we consider a discrete approximation to the process in which the rod is divided up into small segments, each of length Δ. In setting up the discrete approximation we shall often ignore terms of higher order than Δ since the partial differential equations obtained by letting Δ approach 0 will not depend upon higher order terms. This still leaves considerable freedom in the choice of the discrete approximation. We take advantage of this freedom by selecting a discrete approximation which preserves two important features of the physical problem - nonnegativity and conservation of matter. The difference equations will be chosen so that the components of $r(x,y)$, $t(x,y)$ and $\ell(x,y)$ are all obviously nonnegative and so that

$$\sum_{i=1}^{N} (r_i(x,y) + t_i(x,y) + \ell_i(x,y)) = \sum_{i=1}^{N} y_i. \tag{3}$$

We now consider what happens in the segment $[x,x+\Delta]$. When a flux of u_j particles in state j enters the segment at one end and a flux of v_k particles in state k enters at the other end, then the following effects occur:

(a) The expected number of particles from the u_j stream absorbed by the medium is $f_{jj}(x)u_j\Delta + O(\Delta^2)$, and the expected number absorbed from the v_k stream is $f_{kk}(x)v_k\Delta + O(\Delta^2)$.

(b) The expected number of particles from the u_j flux back-scattered in state i is $b_{ij}(x)u_j\Delta + O(\Delta^2)$, and the expected number from the v_k flux back-scattered in state i is $b_{ik}(x)v_k\Delta + O(\Delta^2)$.

(c) The expected number from the u_j stream transmitted in state $i(i\neq j)$ is $d_{ij}(x)u_j\Delta + O(\Delta^2)$ and the expected number from the v_k flux transmitted in state $i(i\neq k)$ is $d_{ik}(x)v_k\Delta + O(\Delta^2)$.

In addition to these effects, which would occur in a no-interaction model, we introduce collision effects. Let u denote the column vector with jth component u_j for $j=1,2,\ldots,N$ and let v denote the vector with components $v_1,v_2,\ldots,v_N$. Then:

(d) The expected number of particles from the u_j stream annihilated due to interactions with the opposing stream with intensities given by the vector v is

$$u_j\varphi_j(u_j,\ v,\ x)\Delta + O(\Delta^2) = \\ = u_j\left[1 - e^{-\Delta\varphi_j(u_j,v,x)}\right] + O(\Delta^2). \tag{4}$$

Similarly, the expected number annihilated from the v_k stream due to interactions with the opposing streams represented by the vector u is

$$v_k\left[1 - e^{-\Delta\varphi_k(v_k,\ u,\ x)}\right] + O(\Delta^2). \tag{5}$$

The functions φ_j are constrained by the conditions

$$\varphi_j(u_j,\ v,\ x) \geq 0, \quad \varphi_j(u_j,\ 0,\ x) = 0, \\ \lim u_j\varphi_j(u_j,\ v,\ x) = 0 \tag{6}$$

In setting up the difference approximation we have used the exponentials to obtain terms which are obviously nonnegative and uniformly bounded if the conditions of (6) hold.

In the following discussion we shall usually suppress the x-dependence of $f_{jj}(x)$, $b_{ij}(x)$, etc. We assume that

$$b_{ij} \geq 0, \quad d_{ij} \geq 0 \ (i \neq j), \quad f_{jj} \geq 0; \tag{7}$$

and we define d_{jj} by the equation

$$d_{jj} = -\left[\sum_{i=1}^{N} b_{ij} + \sum_{\substack{i=1 \\ i\neq j}}^{N} d_{ij} + f_{jj}\right] \quad (j=1,2,\ldots,N) \tag{8}$$

which implies that $d_{jj} \leq 0$. If we introduce the matrices

$$D(x) = (d_{ij}(x)), \ B(x) = (b_{ij}(x)), \ F(x) = (f_{jj}(x)\delta_{ij}), \tag{9}$$

$$M = \begin{bmatrix} 1 & 1 & \ldots & 1 \\ 0 & 0 & \ldots & 1 \\ & \ldots & & \\ 0 & 0 & \ldots & 0 \end{bmatrix}$$

then from (8) we obtain the relation

$$M(B + D + F) = 0. \tag{10}$$

It is also convenient to introduce the matrix

$$\Phi(u, v, x) = (\varphi_i(u_i, v, x)\delta_{ij}). \tag{11}$$

To derive the difference equations for our discrete approximation we introduce the auxiliary vector function $v(x, y)$ where

$$v_i(x,y) = \text{the expected flux in state } i \text{ moving to the left at } x \text{ due to an incident flux at } x + \Delta \text{ of intensity } y. \tag{12}$$

To terms of order Δ, the flux $v(x, y)$ arises from forward scattering of y diminished by absorption and annihilation, and from back scattering of the flux incident at x which is moving to the right. The flux at x moving to the right is $r(x, v(x, y))$. Hence to terms of order Δ we obtain

$$v(x, y) = (I + \Delta D)e^{-\Delta\Phi(y, r(x, v(x,y)))}y + \\ + \Delta B r(x, v(x,y)), \tag{13}$$

an implicit equation for v. Since this implicit equation seems to be somewhat difficult to handle, we shall replace $\Phi(y, r(x, v(x,y)))$ by $\Phi(y, r(x, y))$. If $r(x, v)$ satisfies a uniform Lipschitz condition and $v(x, y) = y + O(\Delta)$, this replacement will only affect terms of higher order.

The reflected flux at $x + \Delta$ arises from fluxes of intensity y and $r(x, v(x, y))$ incident upon the segment from x to $x + \Delta$. Hence to terms of order Δ we obtain

$$r(x + \Delta, y) = \Delta B y + (I + \Delta D)e^{-\Delta\Phi(r(x,y),y)}r(x,v(x,y)) \tag{14}$$

where the two terms represent backward and forward scattering respectively. Similarly we obtain

$$t(x + \Delta, y) = t(x, v(x, y)). \tag{15}$$

In calculating the loss functions we break it up into losses occurring in the segment $[0,x]$ and in the segment $[x, x + \Delta]$. In the latter we must consider losses due to absorption from both the y flux and the $r(x, v(x, y))$ flux, and losses due to annihilation of particles from both fluxes. We obtain

$$\ell(x+\Delta, y) = \ell(x, v(x,y)) + \Delta F y + \Delta F r(x, v(x, y)) + \\ + [I + \Delta D]\left[I - e^{-\Delta\Phi(r(x,y),y)}\right]r(x,v(x, y)) + \\ + [I + \Delta D]\left[I - e^{-\Delta\Phi(y,r(x,y))}\right]y. \tag{16}$$

3. THE DEFINING EQUATIONS

In the following discussion we shall proceed rigorously using the difference equations obtained above as our starting point. The fundamental vector equations for $x = 0, \Delta, 2\Delta, 3\Delta, \ldots, y \geq 0$, are thus the difference equations

$$
\begin{aligned}
r(x+\Delta,\ y) &= \Delta By + (I+\Delta D)e^{-\Delta\Phi(r(x,y),y)}r(x,v(x,y)),\\
t(x+\Delta,\ y) &= t(x,\ v(x,\ y)),\\
\ell(x+\Delta,\ y) &= \ell(x,\ v(x,\ y)) + \Delta Fy + \Delta Fr(x,\ v(x,y)) + \\
&\quad + [I + \Delta D]\left[I - e^{-\Delta\Phi(y,r(x,y))}\right]y + \\
&\quad + [I + \Delta D]\left[I - e^{-\Delta\Phi(r(x,y),y)}\right]r(x,\ v(x,y)) +
\end{aligned}
\tag{1}
$$

and the implicit equation

$$
v(x,\ y) = (I + \Delta D)e^{-\Delta\Phi(y,r(x,y))}y+\Delta Br(x,\ v(x,y)). \tag{2}
$$

For intermediate values of x we shall assume that the functions $r(x,\ y)$, $t(x,\ y)$, and $\ell(x,\ y)$ are defined by linear interpolation.

For the problem we have discussed we have the following initial conditions for the rod.

$$
r(0,\ y) = 0,\quad t(0,\ y) = y,\quad \ell(0,\ y) = 0; \tag{3}
$$

but to allow comparison with other initial value problems for partial differential equations we shall allow the more general situation in which

$$
r(0,\ y) = g(y),\quad t(0,\ y) = y,\quad \ell(0,y) = 0. \tag{4}
$$

In our problem this situation would arise if there were a source at 0 or additional material to the left of 0. We shall assume that $g(y) \geq 0$ and satisfies a Lipschitz condition

$$
\|g(y) - g(y')\| \leq L_0 \|y - y'\|, \tag{5}
$$

where L_0 is a positive constant.

The scheme (1) and (2) is not immediately usable for numerical purposes because (2) is an implicit equation for v. For numerical purposes we would expect to solve (2) by iteration starting with the initial approximation

$(I + \Delta D)e^{-\Delta\Phi(y,\ r(x,y))}y$ for v. Observe that if the iterations converge to a solution of (2) then because of (2.7) all components of the limit vector $v(x,y)$ will be nonnegative provided Δ is sufficiently small so that $I + \Delta D$ has nonnegative elements. Similarly, nonnegativity of r, t, and ℓ is preserved.

4. LIMITING DIFFERENTIAL EQUATIONS

From the foregoing equations, we obtain formally a set of nonlinear partial differential equations for r, t, and ℓ by passing to the limit in Δ. Introduce the three Jacobian matrices

$$T_y = \frac{\partial t_i(x,y)}{\partial y_j}, \qquad R_y = \frac{\partial r_i(x,y)}{\partial y_j}$$

$$L_y = \frac{\partial \ell_i(x,y)}{\partial y_j} \tag{1}$$

Using (3.2) in conjunction with (3.1) we readily obtain the nonlinear partial differential equations

$$r_x - R_y[D_y + Br - \Phi(y,r)y] = By + Dr - \Phi(r,y)r,$$

$$t_x - T_y[D_y + Br - \Phi(y,r)y] = 0$$

$$\ell_x - L_y[D_y + Br - \Phi(y,r)y] = Fy + Fr + \Phi(y,r)y + \Phi(r,y)r.$$

From (3.3) we obtain the initial conditions

$$r(0,Y) = g(y), \quad t(0,y) = y, \quad \ell(0,y) = 0. \tag{3}$$

5. CONSERVATION FOR THE DISCRETE APPROXIMATION

For the considerations of this section it is convenient to replace the initial conditions (3.3) by the conditions

$$r(0,y) = z, \quad t(0,y) = y, \quad \ell(0,y) = 0 \tag{1}$$

where z is an arbitrary nonnegative N-dimensional vector which may depend upon y. The discrete approximation given by (3.1) and (3.2) was chosen so that the following conservation relation would hold:

$$M[r(x,y) + t(x,y) + \ell(x,y)] = M[y + z]. \tag{2}$$

This states that the total reflected flux, transmitted flux, and dissipation due to absorption and annihilation must equal the total input at the ends of the rod - what goes in must equal what goes out.

We shall take up the question of existence of solutions of (3.1) and (3.2) later. In this section, assuming the existence of solutions of (3.1) and (3.2), we prove by induction that the conservation relation (2) holds for $x = 0, \Delta, 2\Delta, 3\Delta, \ldots$. First, for $x=0$ the conservation relation follows immediately from (1). As our induction hypothesis we assume that (2) holds for x for all input vectors y and z, and we then show that it holds for $x + \Delta$. We have, abbreviating $r(x,y)$ by r and $v(x,y)$ by v and making use of (2.10).

$$\begin{aligned}
&M[r(x+\Delta, y)+t(x+\Delta, y)+\ell(x+\Delta, y)] \\
&\quad = \Delta MBy+Mr(x, v)+\Delta MDr(x, v)-(M[I+\Delta D]\left[I-e^{-\Delta\Phi(r,y)}\right]r(x,v)+ \\
&\qquad + Mt(x, v)+M\ell(x, v)+\Delta MFy+\Delta MFr(x, v) + \\
&\qquad + M[I+\Delta D]\left[I-e^{-\Delta\Phi(y,r)}\right]y+M[I+\Delta D]\left[I-e^{-\Delta\Phi(r,y)}\right]r(x,v) \\
&\quad = M[r(x, v)+t(x, v)+\ell(x, v)] + \Delta M[B+F]y + \\
&\qquad + \Delta M[D+F]r(x,v)+M[I+\Delta D]\left[I-e^{-\Delta\Phi(y,r)}\right]y \\
&\quad = M[v+z] - \Delta MDy - \Delta MBr(x, v)+M[I+\Delta D]y - \\
&\qquad - M[I+\Delta D]e^{-\Delta\Phi(y,r)}y = M[y+z],
\end{aligned}$$

which completes the proof by induction.

6. EXISTENCE OF SOLUTIONS FOR DISCRETE APPROXIMATION

We shall now prove the existence of a solution of (3.1)-(3.2) satisfying the initial conditions (3.3) for a small x interval provided the step size Δ is sufficiently small. At the same time we shall show that $r(x,y)$ satisfies a Lipschitz condition with respect to y which is uniform in Δ. In addition to the assumptions already made concerning $B(x)$ are uniformly bounded and $\Phi(u, v, x)$ satisfies uniform Lipschitz conditions with respect to u and v.

The proof will be carried out inductively for a region of the form

$$y \geq 0, \quad \| y \| \leq ce^{-\gamma x} - c_1 \tag{1}$$

where γ is any positive constant for which

$$y > \| B \| + \| D \|, \tag{2}$$

c is an arbitrary positive constant, and $c_1 > \| z \|$, where $z = g(y)$ is the input flux vector at 0. We show that the system (3.1)-(3.2) has a solution for $0 \leq x \leq x^*$ provided $0 < \Delta \leq \delta$, where x^* and δ depend upon the constants c, y, and c_1 as well as upon $B(x)$, $D(x)$, $\Phi(u, v, x)$ and $g(y)$.

As our induction hypothesis we assume that for y and y' satisfying (1)

$$\| r(x,y) - r(x,y') \| \leq L(x) \| y - y' \| \tag{3}$$

where $L(x)$ is a certain function of x to be specified below, such that the conservation relation

$$M[r(x,y) + t(x,y) + \ell(x,y)] = M(y + z) \tag{4}$$

holds, and that $r(x,y) \geq 0$. Under these assumptions we prove that for $0 < \Delta \leq \delta$, $0 \leq x \leq x^*$ the equation (3.2) has a nonnegative solution $v(x,y)$ for $y \geq 0$, $\| y \| \leq c \exp(-y(x+\Delta)) - c_1$. It will then follow immediately that for $y \geq 0$,

$$\| y \| \leq c\, e^{\gamma(x+\Delta)} - c_1 \tag{5}$$

the vectors $r(x+\Delta, y)$, $t(x+\Delta, y)$, and $\ell(x+\Delta, y)$ determined by (3.1) are nonnegative and by the proof given previously satisfy $M[\gamma(x+\Delta,y) + t(x+\Delta,y) + \ell(x+\Delta,y)] = M(y+z)$. We then show that for y and y' in the region (6.1)

$$\| r(x+\Delta, y) - r(x+\Delta, y') \| \leq L(x+\Delta) \| y - y' \|, \tag{6}$$

thus completing the inductive step from x to $x+\Delta$.

We obtain a solution of (3.2) by iteration starting with the vector $v_0 = (I+\Delta D)e^{-\Delta\Phi}y$ and defining the vectors $v_1, v_2, \ldots$ recursively by the equation $v_{n+1} = v_0 + \Delta B r(x, v_n)$. Because all elements of the matrix B are nonnegative, each component of v_n increases monotonely with n. Thus to prove that the sequence $v_0, v_1, v_2, \ldots$ converges it is sufficient to show that the sequence is uniformly bounded. By (4) we have

$$\| v_{n+1} \| \leq \| v_0 \| + \| \Delta B \| (\| v_n \| + \| x \|); \tag{7}$$

hence

$$\|v_{n+1}\| + c_1 \le \|v_0\| + c_1 + \|\Delta B\|(\|v_n\| + c_1), \tag{8}$$

and by induction

$$\|v_{n+1}\| + c_1 \le (\|v_0\| + c_1)(1 + \|\Delta B\| + \|\Delta B\|^2 + \\ + \cdots + \|\Delta B\|^n) \tag{9}$$

Thus if $\Delta\|B\| < 1$ the sequence converges and the limit function $v(x, y) = \lim_{n\to\infty} v_n$ satisfies

$$\|v(x, y)\| + c_1 \le \frac{\|v_0\| + c_1}{1 - \Delta\|B\|} \le \frac{1 + \Delta\|D\|}{1 - \Delta\|B\|}(\|y\| + c_1)$$

$$< (1+\gamma\Delta)(\|y\| + c_1)$$

$$< e^{\gamma\Delta}(\|y\| + c_1) \tag{10}$$

provided Δ is sufficiently small. Note that $v(x, y) \ge 0$ and that if y satisfies (1), then $\|v(x, y)\| \le c \exp(-\gamma(x+\Delta))-c_1$. Because $r(x, v)$ is a continuous function of v in this region, it follows that the limit function $v(x, y)$ satisfies the equation (3.2).

To establish that $r(x + \Delta, y)$ satisfies a Lipschitz condition, we first prove that $v(x, y)$ satisfies a Lipschitz condition for y and y' in the region (5). We have

$$v(x, y) = y+\Delta De^{-\Delta\Phi(y,\ r(x,\ y))}y+(e^{-\Delta\Phi(y,\ r(x,\ y)}-I)y+ \\ + \Delta Br(x, v(x, y)). \tag{11}$$

It is easily seen that if $L(x) \ge L_0$ there is a positive constant c_2 such that

$$\begin{aligned} &\|\Delta De^{-\Delta\Phi(y,\ r(x,\ y))}y - \Delta De^{-\Delta\Phi(y',\ r(x,\ y'))}y'\| \\ &\le c_2 L(x)\Delta\|y-y'\|, \\ &\|(e^{-\Delta\Phi(y,\ r(x,\ y))} - I)y-(e^{-\Delta\Phi(y',\ r(x,\ y'))} - I)y'\| \\ &\le c_2 L(x)\Delta\|y-y'\|. \end{aligned} \tag{12}$$

If y and y' satisfy (1) then by (10)

$$\|v(x, y)\| \le ce^{-\gamma x}-c_1, \ \|v(x, y)\| \le ce^{-\gamma x}-c_1. \tag{13}$$

Hence

$$\begin{aligned}\|v(x,y)-v(x,y')\| &\le \|y-y'\| + 2c_2\Delta L(x)\|y-y'\| + \\ &+ \Delta\|B\| \ \|r(x,v(x,y))-r(x,v(x,y'))\| \\ &\le \|y-y'\|(1+2c_2\Delta L(x)) + \Delta\|B\|L(x)\|v(x,y)-v(x,y')\|,\end{aligned} \tag{14}$$

and consequently

$$\|v(x,y)-v(x,y')\| \le \frac{1+2c_2\Delta L(x)}{1-\Delta\|B\|L(x)}\|y-y'\| < (1+c_3\Delta L(x))\|y-y'\| \tag{15}$$

if $\Delta\|B\|L(x) < 1/2$, where we can let $c_3 = 2\|B\| + 4c_2$.

We have

$$\begin{aligned}r(x+\Delta, y) &= \Delta By+r(x,v(x,y)) +\Delta De^{-\Delta\Phi(r(x,y),y)}r(x,v(x,y)+ \\ &+ (e^{-\Delta\Phi(r(x,y),y)}-I)r(x,v(x,y)).\end{aligned} \tag{16}$$

For y and y' satisfying (1) using (15) we obtain the estimates

$$\begin{aligned}&\|\Delta De^{-\Delta\Phi(r(x,y),y)}r(x,v(x,y)) -\Delta De^{-\Delta\Phi(r(x,y'),y')}r(x,v(x,y'))\| \\ &\qquad \le c_4\Delta L(x)(1+c_3\Delta L(x))\|y-y'\|,\end{aligned} \tag{17}$$

$$\begin{aligned}&\|(e^{-\Delta\Phi(r(x,y),y)}-I)r(x,v(x,y)) -(e^{-\Delta\Phi(r(x,y'),y')}-I)r(x,v(x,y'))\| \\ &\qquad \le c_4\Delta L(x)(I+c_3\Delta L(x))\|y-y'\|\end{aligned}$$

where c_4 is a positive constant. It then follows that

$$\begin{aligned}\|r(x+\Delta, y)-r(x+\Delta, y')\| &\le \|B\| \ \|y-y'\|+L(x)\|v(x, y)-v(x,y')\|+ \\ &+ 2c_4\Delta L(x)(1+c_3\Delta L(x))\|y-y'\| \\ &\le \{\Delta\|B\|+L(x)[1+(c_3+2c_3c_4)\Delta L(x)+2c_4\Delta]\}\|y-y'\|.\end{aligned}$$

Consequently, (6) follows if we define $L(x+\Delta)$ by the equation

$$L(x+\Delta) = c_5\Delta+L(x)(1+2c_5\Delta+c_5\Delta L(x))$$

with $c_5 = \max(\|B\|, c_3+2c_3c_4, c_4)$.

From a well-known lemma

$$L(x) \le \left[\frac{I}{I+L(0)} - c_5 x\right]^{-1} -1$$

if

$$c_5 x \le 1/(1+L(0)).$$

In view of (3.4) the induction can be started by taking $L(0) = L_0$ and can be continued in the region (1) as long as the condition $\Delta\|B\|L(x) < 1/2$ is satisfied. By restricting x to a small enough interval we obtain from (3) the result that r(x, y) satisfies a Lipschitz condition

$$\|r(x, y) - r(x, y')\| \le K\|y-y'\|$$

where K is independent of x and Δ.

7. CONSERVATION FOR NONLINEAR EQUATIONS

Invariant imbedding is a mathematical theory designed to handle a variety of conceptual, analytic, and computational aspects of mathematical physics in a unified fashion without the intervention of boundary-value problems. By means of appropriate choices of space and time variables, all problems are of initial-value type.

Invariant imbedding is a systematic application and extension of the "invariance principles" introduced into the study of radiative transfer by Ambarzumian and Chandrasekhar. More generally, it utilizes the "point-of-regeneration" technique of the type used in the study of branching processes.

In collaboration with Ueno, we have derived a large number of functional equations describing a variety of physical processes, and carried out some large-scale numerical calculations.

8. THE MATRIX RICCATI EQUATION

The equation we wish to study is

$$R'(x) = B(x) + D(x)R(x) + R(x)D(x) + R(x)B(x)R(x),$$

$$R(0) = 0, \tag{1}$$

where B, D, and R are N × N matrices and it is assumed that

$$(a)\ d_{ij}(x) \geq 0, \quad i \neq j,$$
$$(b)\ d_{jj}(x) \leq 0, \tag{2}$$
$$(c)\ b_{ij}(x) \geq 0.$$

Rather than tackle this equation directly, let us indicate its physical source, and then show how the simultaneous consideration of $R(x)$ and two related functions enable us to establish existence of the solution of (1) for all $x > 0$ in a simple and painless fashion.

9. STEADY-STATE NEUTRON TRANSPORT WITH DISCRETE ENERGY LEVELS

Les us begin by describing a model of a steady-state transport process which will be the explicit or implicit source of many of the analytical ideas we shall utilize in what follows.

Consider an idealized neutron-transport process taking place in a one-dimensional, homogeneous, isotropic rod extending along an axis from $z = 0$ to $z = x$. We suppose initially that there are only a finite number, N, of different types of particles moving along the rod. These possible states can be considered to be energy levels, labelled $j = 1, 2, \ldots, N$.

It is assumed that when a particle in state i transverses a segment of the rod, it is subject to interactions with the substance composing the rod. These interactions produce two possible effects: forward or backward scattering into any of the N possible states, and absorptions. However, no fission occurs, which is to say, there is no spontaneous generation of new particles.

It follows that the total number of particles in the process, taking account of those absorbed as well as of those scattered, is changed only by addition from an external source. This obvious conservation principle will be the key to the result obtained below concerning the existence of the solutions of (8.1).

We shall exclude the possibility of collisions or interactions between neutrons themselves. This will permit us to use ordinary differential equations in our application of invariant imbedding.

Finally, let us note that as far as the analysis is concerned, one-dimensional transport with energy levels is equivalent to two-dimensional transport in a plane parallel slab with energy and angular dependence. We are thus treating a quite general transport process connected with a geometric figure such as sphere, cylinder, or plane-parallel region.

10. ANALYTIC PRELIMINARIES

Let us now make the model of a transport process discussed above more precise. We suppose that when a particle in state j ($j = 1, 2, \ldots, N$) enters the infinitesimal segment of length Δ contained in $[x+\Delta, x]$ from either direction (the assumption of isotropy), the following events take place:

(a) The expected number leaving the segment in state j, moving in the same direction, is $1+d_{jj}(x)\Delta + O(\Delta^2)$.

(b) The expected number leaving the segment in state i, $i \neq j$, moving in the same direction is, $d_{ij}(x)\Delta + O(\Delta^2)$. (Forward scattering)

(c) The expected number leaving the segment, moving the opposite direction in state i, is $b_{ij}(x)\Delta + O(\Delta^2)$. (Back scattering.)

(d) The expected number absorbed by the medium is

$$f_{jj}(\Delta) + O(\Delta^2). \tag{1}$$

We call the matrices $D(x) = (d_{ij}(x))$, $B(x) = (b_{ij}(x))$, $F(x) = (f_{ii}(x)\delta_{ij})$, the forward scattering, back scattering, and absorption matrices, respectively.

We assume, on physical grounds, that

$$\begin{aligned} &(a)\ d_{ij}(x) \geq 0 \quad i \neq j \\ &(b)\ b_{ij}(x) \geq 0, \\ &(c)\ f_{ii}(x) \geq 0, \end{aligned} \tag{2}$$

for $x \geq 0$. The basic assumption of the conservation of matter requires that

$$d_{jj}(x) = -\left[\sum_{i=1}^{N} b_{ij}(x) + \sum_{\substack{i=1 \\ i\neq j}}^{N} d_{ij}(x)+f_{jj}(x)\right], \quad j=1,2,\ldots,N. \tag{3}$$

This implies that $d_{jj}(x) \leq 0$, a condition required to account for the increments to other states and for the particles absorbed.

Including the matrix

$$M = \begin{bmatrix} 1 & 1 & \dots & 1 \\ 0 & 0 & \dots & 0 \\ \cdot & \cdot & & \\ \cdot & \cdot & & \\ \cdot & \cdot & & \\ 0 & 0 & \dots & 0 \end{bmatrix}, \tag{4}$$

The relations of (3) can be written in the simple form

$$M(B(x) + D(x) + F(x)) = 0, \quad x \geq 0. \tag{5}$$

This is the fundamental conservation assumption which will yield a corresponding conservation relation for the matrix functions introduced.

11. REFLECTIONS, TRANSMISSION, AND LOSS MATRICES

Let us now introduce the following functions
For i, j = 1, 2, ..., N, let

$r_{ij}(x)$ = expected flux of the neutrons in state i, reflected from a rod of length x, resulting from an incident flux at x of unit intensity in state j;

$t_{ij}(x)$ = expected flux of neutrons in state i, transmitted through a rod of length x resulting from an incident flux at x of unit intensity in state j;

$\ell_{ij}(x)$ = expected flux of neutrons in state i, absorbed within a rod of length x, resulting from an incident flux at x of unit intensity in state j. (1)

Schematically,

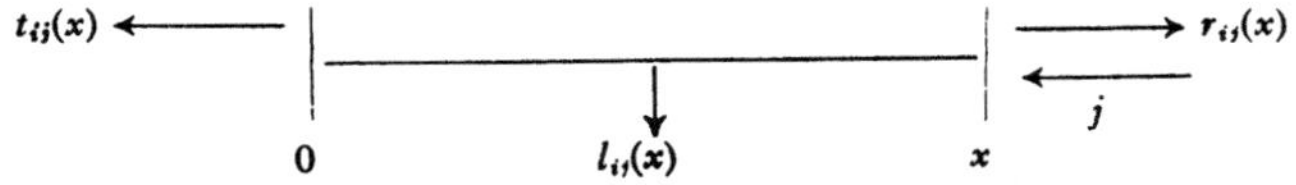

Figure 1

When we say a rod of length x, we mean one whose ends are respectively at the fixed position 0 and the variable position x, as pictured above. As a consequence of the assumption made above concerning no interaction between neutrons, the reflections, transmissions and absorptions depend linearly upon the intensity of the incident flux. Hence we may restrict ourselves here to unit incident fluxes.

Let $R(x) = (r_{ij}(x))$, $T(x) = t_{ij}(x))$, $L(x) = (i_{ij}(x))$ be called respectively the reflection, transmission, and absorption matrices. Using invariant imbedding techniques, we can derive differential equations for these matrices. For the sake of completeness, let us present the derivation here.

Consider the process described above for a rod of length $x + \Delta$ and let an incident flux c be applied at $x + \Delta$. Here c is a vector flux whose j-th component represents the intensity of the incident flux in state j. By virtue of the assumptions of Sec. 10, this results in a flux of $(I + D\Delta)c$ incident at x, a flux of $Bc\Delta$ reflected from $x + \Delta$, and a flux of $Fc\Delta$ absorbed, all to terms in $O(\Delta^2)$. The flux $(I + D\Delta)c$ incident at x results in a flux $R(x)\ (I + D\Delta)c$ reflected at x, a flux $T(x)(I + D\Delta)c$ transmitted through the rod, and a flux $L(x)(I + D\Delta)c$ absorbed.

The flux $R(x)(I + D\Delta)c$ now enters the segment $[x, x+\Delta]$ and results in further interactions. As a result of this, we have the additional reflection $(1 + D\Delta)R(x)(I + \Delta D)c$, an additional absorption

$$(\Delta F)R(x)(I + D\Delta)c = \Delta FR(x)c + O(\Delta^2),$$

and a quantity $\Delta BR(x)c + O(\Delta^2)$ as incident flux upon $[x, 0]$. This incident flux results in a reflection from $[x, 0]$ of $R(x)(\Delta BR(x)c + O(\Delta^2)$, a transmitted flux of $T(x)(\Delta BR(x)c) + O(\Delta^2)$, and a loss of $L(x)(\Delta BR(x)c) + O(\Delta^2)$. The flux $RBR\Delta + O(\Delta^2)$ through $[x, x+\Delta]$ contributes $RBR\Delta + O(\Delta^2)$ to the total reflected flux.

Schematically

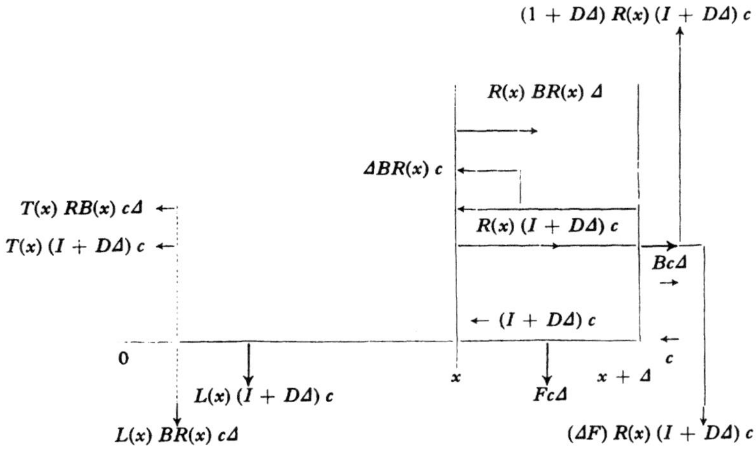

Figure 2

Adding up these effects, we obtain the recurrence relations

$$
\begin{aligned}
R(x+\Delta)c &= Bc\Delta+(I+D\Delta)R(x)(I+D\Delta)c+R(x)BR(x)c\Delta+O(\Delta^2),\\
T(x+\Delta)c &= T(x)(I+D\Delta)c+T(x)BR(x)c\Delta+O(\Delta^2), \qquad (2)\\
L(x+\Delta)c &= Fc\Delta+L(x)(I+D\Delta)c+FR(x)c\Delta+L(x)BR(x)c\Delta+O(\Delta^2).
\end{aligned}
$$

Since these equations hold for arbitrary c, we can discard c. Expanding the left-hand side, and passing to the limit as $\Delta \to 0$, we obtain the Riccati differential equations

$$
\begin{aligned}
R'(x) &= B+DR(x)+R(x)D+R(x)BR(x),\\
T'(x) &= T(x)(D+BR(x)), \qquad (3)\\
L'(x) &= L(x)(D+BR(x))+F(I+R(x)),
\end{aligned}
$$

with the physically obvious initial conditions

$$R(0) = 0, \qquad T(0) = I, \qquad L(0) = 0. \qquad (4)$$

Observe that of the three functions, it is only the reflection function which occurs alone, independently of the other two. The remaining two functions have been deliberately introduced to take advantage of conservation properties.

12. EXISTENCE AND UNIQUENESS OF SOLUTIONS

The conventional existence and uniqueness theory of ordinary differential equations establishes the existence and uniqueness of a solution of (11.3) over some initial interval $[0, a]$. Since it is intuitively clear that the reflection, transmission, and loss functions must exist for all $x \geq 0$ (since no fission is allowed), the question arises as to how to establish this analytically. For the case of constant coefficients (homogeneous rod), we can reduce the equations to linear equations with constant coefficients and use the explicit solutions to help us. For inhomogeneous equations, this approach is more difficult.

To obtain an equivalent linear equation, let us first proceed formally. Consider the two first order matrix equations

$$\begin{aligned} X' &= EX + FY, \\ Y' &= GX + HY, \end{aligned} \tag{1}$$

where E, F, G, and H can be dependent on x. Consider the matrix $Z = XY^{-1}$. We have

$$\begin{aligned} Z' &= (XY^{-1})' \\ &= (EX+FY)Y^{-1} - X(Y^{-1}(GX+HY)Y^{-1}) \\ &= EZ + F - ZGZ - ZH, \end{aligned} \tag{2}$$

an equation similiar to that satisfied by $R(x)$. The identification is complete if we set

$$E = D, \quad F = B, \quad H = -D, \quad G = -B, \tag{3}$$

so that (1) becomes

$$\begin{aligned} X' &= DX + BY, \\ Y' &= -BX - DY. \end{aligned} \tag{4}$$

This procedure is much more than formal, since it turns out that the equations of (4) are the transport equations obtained by applying the usual procedure to the study of the fluxes inside the rod.

As indicated above, it is not a trivial matter to study (4) when B and D are variable matrices.

To establish nonlocal existence of the solutions of (11.3), we add two ingredients: nonnegativity of the matrices R(x), T(x), and L(x), and the conservation relation

$$M(R(x) + T(x) + L(x)) = M.$$

Both of these conditions are intuitively clear, and, as we shall see, readily established rigorously. Once we have done this, it follows that R(x), T(x), and L(x) are uniformly bounded over any interval of existence. It follows that the solutions can be continued for all $x \geq 0$. We are going through this in some detail since the same line of reasoning can be employed for many classes of functional equations arising in mathematical physics.

13. PROOF OF CONSERVATION RELATION

To establish the conservation relation of (12.1), we consider the function

$$Q(x) = M(R(x) + T(x) + L(x)) \tag{1}$$

and differentiate it with respect to x. We have

$$\begin{aligned} Q' &= M(R' + T' + L') \\ &= M(R + T + L)(BR + D) + M(DR + FR + B + F) \\ &= Q(BR + D) + M(DR + FR + B + F), \end{aligned} \tag{2}$$

upon using the equations of (11.3).

Considered as a differéntial equation in Q, we observe that (2) is satisfied by Q(x) = M, since

$$M(BR + D) + M(DR + FR + B + F) = M(B + D + F)(R + I) = 0, \tag{3}$$

by virtue of (10.5). Since Q(0) = M, we see that Q(x) = M within the interval of existence of R(x), T(x), and L(x). This argument can now be repeated from interval to interval.

It is remarkable that one has to use this sophistication to establish a relation which is so immediate from physical considerations. One would expect in place of (2) merely the relation Q' = 0.

14. PROOF OF NONNEGATIVITY

Local existence and nonnegativity of solutions can be established in several different ways. One way is to convert the original system of differential equations into a set of integral equations. Let us begin by writing the differential equations in the form

$$\frac{d}{dx}(e^{-Dx}R(x)e^{-Dx}) = e^{-Dx}[B+R(x)BR(x)]e^{-Dx}, \quad R(x_0)=R_0,$$

$$\frac{d}{dx}(T(x)e^{-Dx}) = T(x)BR(x)e^{-Dx}, \quad T(x_0)=T_0, \tag{1}$$

$$\frac{d}{dx}(L(x)e^{-Dx}) = [L(x)BR(x)+F+FR(x)]e^{-Dx}, \quad L(x_0)=L_0.$$

Thus an appropriate set of integral equations is

$$\begin{aligned} R(x) &= e^{D(x-x_0)}R_0e^{D(x-x_0)} + \int_{x_0}^{x} e^{D(x-x_1)}[B+R(x_1)BR(x_1)]e^{D(x-x_1)}dx_1 \\ &= \theta_1(R, T, L), \\ T(x) &= T_0e^{D(x-x_0)} + \int_{x_0}^{x} T(x_1)BR(x_1)e^{D(x-x_1)}dx_1 \\ &= \theta_2(R, T, L), \\ L(x) &= L_0e^{D(x-x_0)} + \int_{x_0}^{x} [L(x_1)BR(x_1)+F+FR(x_1)]e^{D(x-x_1)}dx_1 \\ &= \theta_3(R, T, L). \end{aligned} \tag{2}$$

The principal result we wish to employ to establish nonnegativity is that $d_{ij} \geq 0$ implies that e^{Dx} is a nonnegative matrix for $x \geq 0$.

Consider the space S of triples of continuous matrix functions $R(x)$, $T(x)$, and $L(x)$ defined on $x_0 \leq x \leq x_0 + a$, with the initial value $R(x_0) = R_0$, $T(x_0) = T_0$, $L(x_0) = L_0$, all nonnegative matrices, satisfying the constraints

$$\|R(x)\| \le c_1, \quad \|T(x)\| \le c_1, \quad \|L(x)\| \le c_1, \tag{3}$$

where

$$c_1 > \mathrm{Max}[\|R_0\|, \|T_0\|, \|L_0\|]. \tag{4}$$

Consider the mapping θ defined on S by means of the right-hand sides of (2). It is readily seen that T is a contractive mapping of S into itself, provided that a is sufficiently small. Thus, by virtue of the Cacciopoli fixed-point theorem, θ has a unique fixed point, the solution of (2).

Alternatively, we can construct the solutions as the limit of a sequence of successive approximations given by

$$\begin{aligned} R_{n+1} &= \theta_1(R_n, T_n, L_n), \quad n \ge 0 \\ T_{n+1} &= \theta_2(R_n, T_n, L_n), \\ L_{n+1} &= \theta_3(R_n, T_n, L_n). \end{aligned} \tag{5}$$

Applying the foregoing result with $x_0 = 0$, $c_1 > N$, where N is the dimension of the system, we obtain a solution over an interval $0 \le x \le a$. From the conservation relation combined with the nonnegativity of $R(x)$, $T(x)$, $L(x)$ on $0 \le x \le a$, it follows that $R(x)$, $T(x)$, $L(x)$ are uniformly bounded. In fact

$$\|R(x)\|, \quad \|T(x)\|, \quad \|L(x)\| \le N, \quad 0 \le x \le a. \tag{6}$$

We can therefore apply the result with $x_0 = a$ and the same c_1 as before. The solution can thus be continued indefinitely.

A third approach starts with the difference equations obtained from (11.2) by neglecting the terms which are $O(\Delta^2)$ The matrices $R(x)$, $T(x)$, $L(x)$ are defined in this way for $x = 0, \Delta, 2\Delta, \ldots$, and defined by means of linear interpolation for other values of x. Since $I + D\Delta \ge 0$ for small Δ, we see that the matrices obtained in this fashion are nonnegative for $x \ge 0$. As is well known, these functions approach the solutions of the differential equation in an initial interval $0 \le x \le b$, thus once again establishing nonnegativity.

15. STATEMENT OF RESULT

We have thus established the following result.

THEOREM. If

(a) $b_{ij}(x) \geq 0$

(b) $d_{ij}(x) \geq 0, \quad i \neq j,$ (1)

(c) $f_{ij}(x) \geq 0,$

and

$$M(B(x) + D(x) + F(x)) = 0$$

for $x \geq 0$, where $B = (b_{ij}(x))$, $D = (d_{ij}(x))$, $F = (f_{jj}(x)\delta_{ij})$, and $M = (\delta_{ij})$, then the equations

$$\begin{aligned} R'(x) &= B+DR(x)+R(x)D+R(x)BR(x), & R(0) &= 0, \\ T'(x) &= T(x)(D+BR(x)), & T(0) &= I, \\ L'(x) &= L(x)(D+BR(x))+F(I+R(x)), & L(0) &= 0, \end{aligned} \tag{2}$$

possess a unique solution for $x \geq 0$. This solution satisfies the conservation relation

$$M(R(x) + T(x) + L(x)) = M, \tag{3}$$

for $x \geq 0$.

In physical terms, this means that the reflection, transmission, and loss matrices are defined for $x \geq 0$, and satisfy the equations of invariant imbedding.

BIBLIOGRAPHY AND COMMENTS

Section 1. We are following the paper:

R. Bellman and S. Lehman, 'Invariant Imbedding, Particle Interaction and Conservation Relations', Journal of Mathematical Analysis and Applications 10 (1965), 112-122.

For other results in the theory of invariant imbedding, see:

R. Bellman and G.M. Wing, An Introduction to Invariant Imbedding, John Wiley & Sons, Inc., New York, 1974.

Relative invariants also play an important role in consideration of nonlinear equations and equations with stochastic terms. See:

R. Bellman, Selective Computations, to appear.

Section 7-15 are largely based on the following paper:

R. Bellman et al. 'Existence and Uniqueness Theorems in Invariant Imbedding. I: Conservation Principles', Journal of Mathematical Analysis and Application 10 (1965), 234-243.

Chapter III

DYNAMIC PROGRAMMING AND PARTIAL DIFFERENTIAL EQUATIONS

1. INTRODUCTION

In this chapter, we wish to show that dynamic programming applied to the calculus of variations leads to various classes of partial differential equations. In the following chapter we will discuss this further.

In Section 7, we show how this new approach yields upper and lower bounds for the solution for nonlinear partial differential equations.

2. CALCULUS OF VARIATIONS AS A MULTISTAGE DECISION PROCESS

So far we have considered that decisions were made at discrete times, a realistic assumption. It is, however, also of considerable interest to examine the fiction of continuous decision processes. In particular, we wish to demonstrate that we can profitably regard the calculus of variations as an example of a multistage decision process of continuous type. In consequence of this, dynamic programming provides a number of new conceptual, analytic, and computational approaches to classical and modern variational problems, particularly to those arising in control processes.

To illustrate the basic idea, which is both quite simple and natural from the standpoint of a control process, let us consider the scalar functional

$$J(u) = \int_0^T g(u, u')dt \tag{1}$$

The problem of minimizing J(u) over all u satisfying the initial condition u(0) = c leads along classical lines as we know to the task of solving the Euler equation

$$\frac{\partial g}{\partial u} - \frac{d}{dt}\left(\frac{\partial g}{\partial u}\right) = 0 \tag{2}$$

subject to the two-point condition

$$u(0) = c, \quad \left.\frac{\partial g}{\partial u}\right|_{t = T} = 0. \tag{3}$$

This equation, as we have already seen, is a variational equation obtained by considering the behavior of the functional $J(u + w)$ for "all" small w where u is the desired minimizing function. This procedure of examining the neighborhood of the extremal in function space is a natural generalization of that used in calculus in the finite-dimensional case. As in the finite-dimensional case, there can be considerable difficulty first in solving the variational equation and then in distinguishing the absolute minimum from other stationary points.

Let us now pursue an entirely different approach motivated by the theory of dynamic programming. In particular, it is suggested by the applications of dynamic programming to the study of deterministic control processes. Rather than thinking of a curve $u(t)$ as a locus of points, let us take it to be an envelope of tangents. Ordinarily, we determine a point on the curve by the coordinates $(t, u(t))$. However, we can equally well trace out the curve by providing a rule for determining the slope u' at each point (t, u) along the path. The determination of the minimizing curve $u(t)$ can thus be regarded as a multistage decision process in which it is necessary to choose a direction at each point along the path. Motivation of this approach in the domain of pursuit processes is easily seen, or equivalently, in the determination of geodesics, in analysis of multistage investment processes, or in the study of optimal growth processes in mathematical economics.

3. A NEW FORMALISM

Let us use the foregoing concept of a minimization process to obtain a new analytic approach to variational problems. See figure 1.

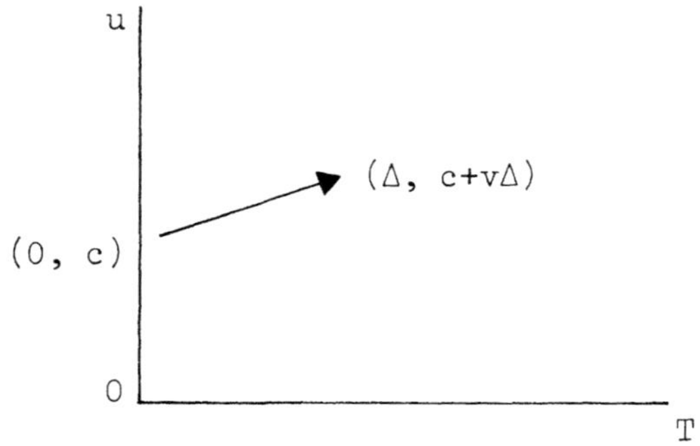

Figure 1

Let us denote the minimum value of J(u) as defined by (2.1) (assumed to exist) by f(c, T). Thus, we introduce the function

$$f(c, T) = \min_u J(u), \tag{1}$$

defined for $T \geq 0$ and $-\infty < c < \infty$. Since we are now interested in policies, the initial state must be considered to be a variable.

For the problem (2.1), a suitable initial condition is $f(c, 0) = 0$. The situation is a bit more complicated when there is a two-point boundary condition.

Let us first proceed formally (as in the derivation of the Euler equation) supposing that all partial derivatives exist, that all limiting procedures are valid, etc.

Let $v = v(c, T)$, as in the foregoing figure, denote the initial direction; $v = u'(0)$, which clearly depends on c and T. This is the missing initial value in (2.2), an essential quantity. Writing

$$\int_0^T = \int_0^\Delta + \int_\Delta^T , \tag{2}$$

a convenient separability property of the integral, we see that for any initial v(c, T) we have

$$f(c, T) = \Delta g(c, v) + O(\Delta^2) + \int_\Delta^T . \tag{3}$$

Here Δ is infinitesimal. We now argue as follows. Regardless of how $v(c, T)$ was chosen, we are going to proceed from the new point $(\Delta, c+v\Delta)$ so as to minimize the remaining integral $\int_\Delta^T g(u, u')dt$. But

$$\min_u \int_\Delta^T g(u, u')dt = f(c + v\Delta, T - \Delta) \tag{4}$$

by definition of the function f. Hence, for any initial choice of v, we have

$$f(c, T) = g(c, v)\Delta + f(c + v\Delta, T - \Delta) + O(\Delta^2). \tag{5}$$

This is an example of the "principle of optimality" for multistage decision processes.

It remains to choose $v(c, T)$ appropriately. Clearly, v should be chosen to minimize the right-hand side of (5). Thus, we obtain the equation

$$f(c, T) = \min_v[g(c, v)\Delta+f(c+v\Delta, T-\Delta)]+O(\Delta^2) \tag{6}$$

Expanding the appropriate terms above in powers of Δ and letting $\Delta \to 0$, we obtain the partial differential equaticn

$$f_T = \min_v[g(c, v) + vf_c]. \tag{7}$$

We have noted above that $f(c, 0) = 0$, an initial condition. Thus we have transformed the original variational problen of minimizing $J(u)$ in (2.1) into that of solving a nonlinear partial differential equation subject to an initial condition.

The foregoing is a cavalier approach in the spirit of the usual first derivation of the Euler equation. We leave it to the reader to spot all the irregularities.

4. LAYERED FUNCTIONALS

Let us call a functional of the form

$$J(u) = \int_0^T g\left(u, u', \int_0^T h(u, u')dt_1\right)dt \tag{1}$$

a layered functional. The problem of minimizing $J(u)$ with respect to u, subject to an initial condition $u(0) = c$, can be approached in the following fashion. Set

$$\int_0^T h(u, u')dt = k \tag{2}$$

where k is a parameter to be determined, and consider the more unusual problem of minimizing the functional

$$J(u, k) = \int_0^T g(u, u', k)dt \tag{3}$$

This determines a function $u(t, k)$. The value of k is to be obtained from the consistency relation

$$\int_0^T h(u(t, k), u'(t, k))dt = k. \tag{4}$$

Clearly, there are some serious analytical and computational obstacles in an approach of this nature.

Let us then examine an alternate approach using dynamic programming. In this fashion we are led to some initial-value problems for nonlinear partial differential equations. Conversely, we are led to some representation theorems for certain classes of nonlinear partial differential equations. These representation theorems can be used to obtain upper, and in some cases lower, bounds for the solutions. In what follows we will present the purely formal aspects.

5. DYNAMIC PROGRAMMING APPROACH

Let us take the more general problem of minimizing

$$J(u, a) = \int_0^T g\left(u, u', a + \int_0^T h(u, u')dt\right)dt \tag{1}$$

subject to $u(0) = c$, where a is a parameter in the range $(-\infty, \infty)$. Introduce the function

$$f(c, a, T) = \min_u J(u, a), \tag{2}$$

and write

$$\varphi(c, a, T) = \int_0^T h(u, u')dt, \tag{3}$$

where u is the function, assumed to exist, which minimizes J(u, a).

Then, proceeding in a familiar fashion, we write

$$f(c, a, T) = \min_u \left[\int_0^\Delta + \int_\Delta^T \right], \tag{4}$$

leading via the principle of optimality to

$$f(c, a, T) = \min_v [g(c, v, a + \varphi)\Delta +$$
$$+ f(c + v, \Delta, a+h(c, v)\Delta, T-\Delta)] + O(\Delta), \tag{5}$$

and thus, in the limit as $\Delta \to 0$, to the partial differential equation

$$fr = \min_v [g(c, v, a + \varphi) + vf_c + h(c, v)f_a]. \tag{6}$$

Similarly, φ satisfies the equation

$$\varphi r = (hc, v) + v\varphi_c + h(c, v)\varphi_a, \tag{7}$$

where $v = v(\varphi, f_c, f_a, fr)$ is determined by (6). The initial conditions are

$$f(c, a, 0) = 0, \quad \varphi(c, a, 0) = 0 \tag{8}$$

We can obtain more general boundary conditions by taking the functional

$$J_1(u, a) = \int_0^T g\left(u, u', a+r(u(T)) +\right.$$
$$\left. + \int_0^T h(u, u')dt\right)dt + s(u(T)) \tag{9}$$

as a starting point. Then (8) is replaced by

$$f(c, a, 0) = s(c), \qquad \varphi(c, a, 0) = r(c). \tag{10}$$

6. QUADRATIC CASE

We can obtain particular classes of quadratically nonlinear partial differential equations in this way by choosing various quadratic functionals such as

$$J_2(u, a) = \int_0^T \left(a + u' + \int_0^T u\, dt\right)^2 dt + \int_0^T u^2 dt. \tag{1}$$

Furthermore, the functional

$$J_3(u, a) = \int_0^T \left[u'^2 + g\left(a + u + \int_0^T u\, dt\right)\right] dt \tag{2}$$

leads to an interesting type of nonlinear partial differential equation.

7. BOUNDS

Since we have a variational problem where we are seeking the minimum, we know that any trial function provides an upper bound. In this way, we readily obtain upper bounds for the solution of the associated partial differential equation.

To obtain lower bounds, we can proceed in several ways. First of all, we can use a dual problem, a procedure introduced by Friedrichs, to obtain lower bounds. Secondly, we can find another problem for the desired quantity which yields a lower bound. Thirdly, we can use geometrical ideas to obtain lower bounds.

BIBLIOGRAPHY AND COMMENTS

Section 1. Many further results and references may be found in:

R. Bellman, Introduction to the Mathematical Theory of Control Processes, Vol. II, Academic Press, Inc., New York, 1971.

Another important technique in dynamic programming approximation in policy space is also useful in studying partial differential equations.

Section 4. We are following the paper:

R. Bellman, 'Functional Equations in the Theory of Dynamic Programming. XV. Layered Functionals and Partial Differential Equations', *Journal of Mathematical Analysis and Applications* 28 (1969), 183.

See also:

R. Bellman, *Methods of Nonlinear Analysis*, Vol. II, Academic Press, New York, 1973.

for a discussion of duality.

Chapter IV

THE EULER-LAGRANGE EQUATIONS AND CHARACTERISTICS

1. INTRODUCTION

In the preceding chapter, we considered the problem of minimizing the functional

$$J(y) = \int_0^T L[x(t), y(t)]\, dt \tag{1}$$

subject to relations of the form

$$\frac{dx_i}{dt} = g_i(x, y), \quad x_i(0) = c_i, \quad i=1, 2, \ldots, n, \tag{2}$$

where x and y represent the n-tuples $(x_1, x_2, \ldots, x_n)$, $(y_1, y_2, \ldots, y_n)$.

Setting

$$f[x(0), T] = \max_y J(y), \tag{3}$$

which is assumed to exist, the principle of optimality yields the functional equation

$$f[x(0), S + T] = \max_{y[0, s]} \left[\int_0^S L(x, y)dt+f(x(S), T)\right], \tag{4}$$

where we minimize over all admissible functions defined over the interval [0, S].

We showed that in the one-dimensional case, the limiting form of Equation (4) as $S \to 0$ yields a partial differential equation for f, and by elimination of f, a partial differential equation for $z = y(0) = y[x(0), T]$, whose characteristics are precisely the Euler-Lagrange equation associated with f.

In this chapter we shall pursue the connection between the

classical approach sketched above, and shall generalize the result concerning the Euler-Lagrange equation and the characteristics of the partial differential equation for $y(0)$.

2. PRELIMINARIES

We shall take L to be a function of the 2n real variables, x_i, y_j, $i,j=1, 2, \ldots, n$, which is twice differentiable with respect to any of the y_j and which is continuously differentiable with respect to any of the variables x_i. The vectors $x(t)$ and $y(t)$ determine respectively the state and the policy of the continuous decision process.

It will be assumed that the Hessian determinant of L with respect to y is nonzero for all x and y, i.e.,

$$L_{yy} = \det L_{y_i y_j} \neq 0 \tag{1}$$

Under the assumption that $\partial(g_1, g_2, \ldots, g_n)/\partial(y_1, y_2, \ldots, y_n) \neq 0$ (the non-vanishing of the Jacobian), the variables y_i may be replaced by the variables g_i, and we may assume, for theoretical purposes, that the constraint of Equation (1.2) has the form

$$\frac{dx_i}{dt} = y_i. \tag{2}$$

Let us now, to preserve covariant and contravariant notation, replace y by the n-tuple $z = (z^1, z^2, \ldots, z^n)$ and x by $(x^1, x^2, \ldots, x^n)$, and write

$$\max_z \int_0^T \bar{L}(x, z_i)dt = f[x(0), T]. \tag{3}$$

Note that $\bar{L}(x, z) = L(x, y)$.

3. THE FUNDAMENTAL RELATIONS OF THE CALCULUS OF VARIATIONS

Returning to the Equation (1.4) (with y replaced by z), we let $S \to 0$. In this way we shall obtain relations between f and L which are known as the transversality conditions. These conditions guarantee that Hilbert's integral is independent of the path of integration. Caratheodory derives these relations by means of a contact transformation and makes them the basis of his treatment of the calculus of variations. Following Caratheodory, we shall call them the fundamental relations of the calculus of variations.

It is interesting to observe that the derivation of these relations obtained from the principle of optimality is distinct from either of the above.

From Equation (1.4) we obtain, the limiting relation

$$\max_{z(0)} \left[\bar{L} + \frac{\partial f}{\partial x^i} z^i(0) - \frac{\partial f}{\partial T} \right] = 0, \tag{1}$$

where we are using the summation convention: i summed from 1 to n. This relation, in turn, implies

$$\begin{aligned} &\bar{L} + \frac{\partial f}{\partial x^i} z^i(0) - \frac{\partial f}{\partial T} = 0, \\ &\frac{\partial \bar{L}}{\partial z^i} + \frac{\partial f}{\partial x^i} = 0. \end{aligned} \tag{2}$$

Now consider the interval $[-t, T]$ of fixed length $t + T = T_0$; translating its left end point to the origin, we make the following definitions:

$$\begin{aligned} q^i(t) &= x^i(0), \\ v^i(t) &= -z^i(0), \\ \phi[q(t), t] &= f[x(0), T]. \end{aligned} \tag{3}$$

Since $dt + dT = 0$, it is clear that $\partial\phi/\partial t = -\partial f/\partial T$ and $v^i(t) = dq^i(t)/dt$. In this notation (3.2) may be solved for $\partial\phi/\partial t$ and $\partial\phi/\partial q^i$.

$$\frac{\partial\phi}{\partial t} = \bar{L} - \frac{\partial\bar{L}}{\partial v^i} v^i,$$
$$\frac{\partial\phi}{\partial q^i} = \frac{\partial\bar{L}}{\partial v^i} . \tag{4}$$

These are the fundamental relations of the calculus of variations.

It will be convenient to write these relations in the Hamiltonian form. First we make the changes of variables

$$p_i = \frac{\partial\bar{L}}{\partial v^i}(v, q),$$
$$q^i = q^i \tag{5}$$

which is possible because the condition in (1.4) ensures that Jacobian $\partial(p_1, \ldots, p_n)/\partial(v^1, \ldots, v^n)$ is nonzero. Then (4) becomes just

$$\frac{\partial\phi}{\partial t} + H(p, q) = 0,$$
$$\frac{\partial\phi}{\partial q^i} = p_i, \tag{6}$$

where H is the Hamiltonian $H = p_i v^i - \bar{L}$. Eliminating p_i in (6) gives the Hamilton-Jacobi equation

$$\frac{\partial\phi}{\partial t} + H\left(\frac{\partial\phi}{\partial q}, q\right) = 0. \tag{7}$$

Since H does not contain t explicitly; (10) shows that $\phi = S - Et$, where E is a constant and S is independent of t. Thus Equation (7) becomes the traditional energy equation

$$H\left(\frac{\partial S}{\partial q}, q\right) = E. \tag{8}$$

4. THE VARIATIONAL EQUATIONS

It is well known that the characteristics of the Hamilton-Jacobi equation are just Hamilton's canonical equations:

$$\frac{dq^i}{dt} = \frac{\partial H}{\partial p_i},$$
$$\frac{dp_i}{dt} = \frac{\partial H}{\partial q_i}, \tag{1}$$

which in turn, are transcriptions of the Euler-Lagrange variational equations. In Section 3.8 we shall show that these are also the characteristics of a new system of partial differential equations in the variables z, which we derive from (3.2) by eliminating f. Before deriving the new equations it will be of interest to give a more usual derivation of the Euler-Lagrange equations under the side conditions that $dx^i/dt = z^i(x, y)$. Suppose that the values $x = \bar{x}(t)$, $y = \bar{y}(t)$ are extremals, then set $x(t) = \bar{x}(t) + \xi(t)$, $y(t) = y(t) + \eta(t)$, where $\xi(t)$ and $\eta(t)$ are n-tuples whose components vanish at both $t = 0$ and $t = T$. We assume that $z = z(x, y)$ has continuous derivatives with respect ot x and y. Then, since

$$\frac{dx^i}{dt} = z^i(x,y) = z^i(\bar{x}, \bar{y}) + \frac{\partial z^i}{\partial x^j}\xi^j + \frac{\partial z^i}{\partial y^j}\eta^j + O(\xi, \eta) \tag{2}$$

and

$$\frac{dx^i}{dt} = \frac{d\bar{x}^i}{dt} + \frac{d\xi^i}{dt} = z^i(\bar{x}, \bar{y}) + \frac{d\xi^i}{dt}, \tag{3}$$

we have

$$\frac{d\xi^i}{dt} = \frac{\partial z^i}{\partial x^j}\xi^j + \frac{\partial z^i}{\partial y^j}\eta^j + O(\xi, \eta). \tag{4}$$

In order to set the first variation of the integral $\int_0^T L(x, y)dt$ equal to zero, i.e.,

$$\int_0^T \left(\frac{\partial L}{\partial x^i}\xi^i + \frac{\partial L}{\partial y^j}\eta^j\right)dt = 0, \tag{5}$$

we need an expression for ξ in terms of η; this expression will be derived from Equation (4). Assume that the values of $\eta = \eta(t)$ have been assigned, and let $(a^i_j) = (a^i_j(t))$ be the $\eta \times \eta$ solution of

$$\frac{da^i_{.j}}{dt} = \frac{\partial z^i}{\partial x^k} a^k_{.j} \tag{6}$$

for which $a^i_{.j}(0) = \delta^i_{.j}$; since $\partial z/\partial x$ and $\partial z/\partial y$ are both continuous, the former satisfies a Lipschitz condition with respect to y; hence there exists a unique solution, $(a^i_{.j})$, of Equation (6) which possesses an inverse, $(\alpha^i_{.j}) = (\alpha^i_{.j}(t))$. Returning to Equation (4), the method of variation of paramaters shows that

$$\xi^i(t) = a^i_{.j} \int_0^t \alpha^j_k \frac{\partial z^k}{\partial y^\ell} \eta^\ell \, d\gamma + O(\xi, \eta). \tag{7}$$

Hence the first term in Equation (5) is

$$\int_0^T \frac{\partial L}{\partial x^i} \xi^i dt = \int_0^T \frac{\partial L}{\partial x^i} a^i_{.j}\left(\int_0^t \alpha^j_k \frac{\partial z^k}{\partial y^\ell} \eta^\ell d\gamma\right)dt + O(\xi, \eta)$$

$$= \int_0^T \left(\int_t^T \frac{\partial L}{\partial x^i} a^i_{.j} d\gamma\right)\alpha^j_k \frac{\partial z^k}{\partial y^\ell} \eta^\ell dt + O(\xi, \eta). \tag{8}$$

Since η is arbitrary, the fundamental lemma of the calculus of variations, applied to Equation (5), shows that

$$\left(\int_t^T \frac{\partial L}{\partial x^i} a^i_{.j} \, d\gamma\right) \alpha^j_k \frac{\partial z^k}{\partial y^\ell} + \frac{\partial L}{\partial y^\ell} = 0, \quad \ell = 1, \ldots, n. \tag{9}$$

Let $\partial y^\ell/\partial z^k$ be the inverse of $\partial z^k/\partial y^\ell$ and multiply Equation (10) by $(\partial y^\ell/\partial z^m)a^m_n$ to find

$$\int_t^T \frac{\partial L}{\partial x^i} a^i_{.j} \, d\gamma + L_{z^k} a^k_{.j} = 0, \tag{10}$$

where

$$L_{z^k} = \frac{\partial L}{\partial y^i} \frac{\partial y^i}{\partial z^k} .$$

Differentiating Equation (10) and using Equation (6) yields

$$-\frac{\partial L}{\partial x^i} a^i_{.j} + \left(\frac{d}{dt} L_{z^k}\right) a^k_{.j} + L_{z^k} \frac{d}{dt} a^k_{.j}$$

$$= -\frac{\partial L}{\partial x^i} a^i_{.j} + \left(\frac{\partial}{\partial t} L_{z^i}\right) a^i_{.j} + L_{z^k} \frac{\partial z^k}{\partial x^i} a^i_{.j} = 0, \tag{11}$$

hence

$$\frac{d}{dt} L_{z^i} + L_{z^k} \frac{\partial z^k}{\partial x^i} = \frac{\partial L}{\partial x^i}, \quad i=1, \ldots, n. \tag{12}$$

These are the Euler-Lagrange equations.

A simpler derivation can be given by taking x and z as the independent variables with the side condition

$$\frac{dx}{dt} = z. \tag{13}$$

The corresponding Euler-Lagrange equations are

$$\frac{d}{dt} \frac{\partial \bar{L}}{\partial z^i} = \frac{\partial \bar{L}}{\partial x^i}, \qquad i=1, \ldots, n, \tag{14}$$

where $\bar{L}(x, z) = L(x, y)$. But

$$\frac{\partial \bar{L}}{\partial z^i} = \frac{\partial L}{\partial z^i} [x, y(x, z)] = \frac{\partial L}{\partial y^j} \frac{\partial y^j}{\partial z^i} = L_{z^i} \tag{15}$$

and

$$\begin{aligned} \frac{\partial L}{\partial x^i} &= \frac{\partial \bar{L}}{\partial x^i} [x, z(x, y)] = \frac{\partial \bar{L}}{\partial x^i} + \frac{\partial \bar{L}}{\partial z^k} \frac{\partial z^k}{\partial x^i} \\ &= \frac{\partial \bar{L}}{\partial x^i} + L_{z^k} \frac{\partial z^k}{\partial x^i} \end{aligned} \tag{16}$$

so that Equation (12) is identical to Equation (14).

There is a more convenient way of writing Equation (12). Suppose that $a^i_{.j} = a^i_{.j}(t)$ is the solution of Equation (16). Then

$$\begin{aligned}\frac{d}{dt}(a^i_j L_{z^i}) &= a^i_j \frac{d}{dt} L_{z^i} + \frac{da^k_j}{dt} L_{z^k} \\ &= a^i_j\left[\frac{d}{dt} L_{z^i} + \frac{\partial z^k}{\partial x^i} L_{z^k}\right].\end{aligned} \tag{17}$$

The expression in brackets is the left-hand side of Equation (12). Hence the Euler-Lagrange equations are

$$\frac{d}{dt}(a^i_j L_{z^i}) = a^i_j \frac{\partial L}{\partial x^i}, \tag{18}$$

which is an immediate consequence of Equation (10). Now that the passage to the description in the variables x and y is clear, we return to the variables x and z for the remainder of the chapter.

5. THE EULERIAN DESCRIPTION

Let us now see how the initial tangents to the external curves depend on the initial values. If we restrict the extremal curves to lie in a field, which means that the expression

$$\left(\bar{L} - \frac{\partial \bar{L}}{\partial v^i} v^i\right) dt + \frac{\partial \bar{L}}{\partial v^i} dq^0$$

is to be a total differential $d\phi$ then at $t = 0$ it is clear that the initial values of v cannot be chosen arbitrarily, but must satisfy certain partial differential equations in the variables q. The condition just mentioned is obviously the same as the fundamental relations. For an arbitrary value of t, we suppose that v is a function of q, $v = v(q, t)$, and we cross-differentiate among the last n equations in (3.4) to find

$$\theta_{ij} \equiv \frac{\partial}{\partial q^i} \bar{L}_{v^j} - \frac{\partial}{\partial q^j} L_{v^i} = 0, \quad i, j=1, \ldots, n, \tag{1}$$

where $(\partial/\partial q^i)\bar{L}_{v^j} = \bar{L}_{v^j q^i} + \bar{L}_{v^j v^k}(\partial v^k/\partial q^i)$. The symbols $\partial/\partial q^i$ and $\partial/\partial t$ henceforth are used to indicate differentiations in which $q^1, \ldots, q^n$ and t are the independent variables, with

$v = v(q, t)$, and subscripts are used to indicate differentiations in which q, t, and v are independent. Note that differentiations indicated by subscripts commute with one another, and that the differentiations $\partial/\partial q^i$, $\partial/\partial t$ also commute with one another; but the differentiations indicated by subscripts do not commute with $\partial/\partial q^i$, $\partial/\partial t$.

The natural way to solve (1) is merely to prescribe an arbitrary function $\phi = \phi(t, q^1, \ldots, q^n)$ for any value of t and then to solve the algebraic relations

$$\frac{\partial\phi}{\partial q^i} = \bar{L}_{v^i}(q, v), \quad i=1, \ldots, n, \tag{2}$$

for $v^i = v^i(q, t)$, $i=1, \ldots, n$.

Once a function ϕ has been selected for $t = 0$, the usual method of computing the extremals is merely to integrate the variational equations with the initial values $q^1(0), \ldots, q^n(0), v^1(0), \ldots, v^n(0)$, unless a complete integral of the Hamilton-Jacobi equations happens to be available. An alternative method can be used. We return to (3.4) and cross-differentiate the first member with each of the last n members, giving rise to n partial differential equations in the unknowns v^i,

$$\frac{\partial}{\partial t}\bar{L}_{v^i} = \frac{\partial}{\partial q^i}\bar{L} - \frac{\partial}{\partial q^i}(\bar{L}_{v^j}v^j) = \bar{L}_{q^i} - v^j\frac{\partial\bar{L}_{v^j}}{\partial q^i}$$

$$i=1, \ldots, n, \tag{3}$$

where $\partial/\partial q^i$ indicates a differentiation in which $v = v(q, t)$, and the subscripts indicate differentiations in which q and v are independent. These equations themselves form a system of partial differential equations which one might integrate under the restriction that the initial values of $v = v(q, 0)$ are solutions to (1) for $t = 0$. However, we go one step further and add an appropriate linear combination of the equations of (1) to these equations to obtain the new system,

$$\frac{\partial}{\partial t}\bar{L}_{v^i} = \bar{L}_{q^i} - v^j\frac{\partial L_{v^j}}{\partial q^i} + \lambda_i^{jk}\theta_{kj}. \tag{4}$$

where $\lambda_i^{jk} = \lambda_i^{jk}(q, v, t)$ is some known multiplier. It will turn out if the initial conditions satisfy (1) at t = 0, then the integrals of (4) satisfy (1) everywhere. However, for certain values of λ_{i}^{jk}, one can find properties of (4) which are independent of the initial conditions; for these values we shall make no assumptions about θ_{kj}.

The first task is to prove the assertion that integrals of (4) which satisfy (1) for t = 0 satisfy it everywhere. Since

$$\frac{\partial \bar{L}}{\partial q^i} = \bar{L}_{q^i} + \bar{L}_{v^j} \frac{\partial v^j}{\partial q^i}$$

$$= \frac{\partial}{\partial t} \bar{L}_{v^i} + v^j \frac{\partial \bar{L}_{v^j}}{\partial q^i} + \bar{L}_{v^j} \frac{\partial v^j}{\partial q^i} - \lambda_{i}^{jk}\theta_{kj} \tag{5}$$

$$= \frac{\partial}{\partial t} \bar{L}_{v^i} + \frac{\partial}{\partial q^i} (v^j \bar{L}_{v^j}) - \lambda_{i}^{j\ell}\theta_{\ell j}, \tag{5}$$

we have

$$\frac{\partial^2 \bar{L}}{\partial q^k \partial q^i} = \frac{\partial^2 \bar{L}_{v^i}}{\partial t \partial q^k} + \frac{\partial^2}{\partial q^i \partial q^k} (v^j \bar{L}_{v^j}) - \frac{\partial}{\partial q^k} (\lambda_{i}^{j\ell}\theta_{\ell j}), \tag{6}$$

hence subtraction gives

$$\frac{\partial}{\partial t} \theta_{ik} = \frac{\partial}{\partial q^i} (\lambda_{k}^{j\ell}\theta_{\ell j}) - \frac{\partial}{\partial q^k} (\lambda_{i}^{j\ell}\theta_{\ell j}). \tag{7}$$

To complete the proof, one needs to give a uniqueness theorem for (7), which is not difficult. Details will be omitted.

Now we specialize λ_{i}^{jk} by taking $\lambda v^j \delta_{i}^{k}$ for some parameter λ. If $\lambda = 0$, then Equation (4) is identical to Equation (3) and Equation (7) becomes

$$\frac{\partial}{\partial t} \theta_{ik} = 0, \tag{8}$$

so that θ_{ik} is independent of t.

For $\lambda = 1$, Equation (4) becomes

$$\frac{\partial}{\partial t}\bar{L}_{v^i} + v^j \frac{\partial \bar{L}_{v^j}}{\partial q^j} = \bar{L}_{q^i}, \tag{9}$$

which has the form of Jacobi's simultaneous equations, i.e., it is a system of n linear first-order partial differential equations in the n unknowns, L_{n^i}, such that all differentiations occur in a single direction. It should be noted that the expressions on the left-hand side of Equation (1) are the material derivatives of L_{v^i} in the sense of hydrodynamics.

Thus Equation (9) may be considered as the equation of motion of a fluid whose velocity is v at the point q, the equation being written in terms of the Eulerian description of the fluid. The corresponding Lagrangian description can be given immediately, since no other derivatives, besides the material derivatives just mentioned, enter into the equations. We take $v(q, t) = \partial^* q/\partial^* t$, where the operator $\partial^*/\partial^* t$ indicates a differentiation in which q is allowed to vary as indicated - by definition, its initial values being held fixed. Henceforth, as in this definition, the asterisks will indicate partial differentiations in which t and the initial values $c^i = q^i_{t=0}$ of q^i are taken as independent variables. In this notation, Equation (9) becomes

$$\frac{\partial^*}{\partial^* t}\bar{L}_{v^i} = \bar{L}_{q^i}, \tag{10}$$

which is just the set of variational equations. (It is a coincidence, in this connection with hydrodynamics, that the variational equations happen to be called the Euler-Lagrange equations.)

It is of interest to see how the quantities θ_{ij} vary along the curves $q = q(t)$ as determined by $v(q, t) = \partial^* q/\partial^* t$. Using the identity

$$\frac{\partial}{\partial q^i}\theta_{jk} + \frac{\partial}{\partial q^j}\theta_{ki} + \frac{\partial}{\partial q^k}\theta_{ij} = 0, \tag{11}$$

we find that Equation (7) becomes

$$\frac{\partial^*}{\partial^* t}\,\theta_{ik} = \frac{\partial v^j}{\partial q^i}\,\theta_{kj} - \frac{\partial v^j}{\partial q^k}\,\theta_{ij}. \tag{12}$$

So, although they are not constant along the extremals, the values of θ_{ik} can easily be determined if they are given initially.

6. THE LAGRANGIAN DESCRIPTION

The previous section has been concerned with the Eulerian description of the flow along the extremals. For comparison, we derive the corresponding Lagrangian description of the same flow by means of a similar technique. Note that according to Equation (3.4) these exists a function ϕ such that

$$d\phi = (\bar{L} - \bar{L}_{v^i}\, v^i)dt + \bar{L}_{v^i}\, dq^i. \tag{1}$$

Now, at t=0, we set $q^i = c^i$, and for $t > 0$ we define q^i in terms of the variables $c^1, \ldots, c^n$, and t, by means of

$$\frac{\partial^* q^i}{\partial^* t} = v^i(q, t) \tag{2}$$

Thus

$$dq^i = \frac{\partial^* q^i}{\partial^* c^j}\, dc^j + \frac{\partial^* q^i}{\partial^* t}\, dt = \frac{\partial^* q^i}{\partial^* c^j}\, dc^j + v^i dt. \tag{3}$$

Now Equation (1) may be written as

$$d\phi = \bar{L}dt + \bar{L}_{v^i}\frac{\partial^* q^i}{\partial^* c^j}\, dc^j = \frac{\partial^* \phi}{\partial^* t}\, dt + \frac{\partial^* \phi}{\partial^* c^j}\, dc^j. \tag{4}$$

Using (2), it is clear that cross- diffentiation between the expressions for $\partial^*\phi/\partial^* t$ and $\partial^*\phi/\partial^* c^j$ gives

$$\bar{L}_{q^j}\,\frac{\partial^* q^j}{\partial^* c^i} = \frac{\partial^*}{\partial^* t}\, L_{v^j}\,\frac{\partial^* q^j}{\partial^* c^i}\;, \tag{5}$$

and that cross-differentiation among the $\partial^*\phi/\partial^* c^j$, j=1, ..., n, gives

$$\omega_{ik} \equiv \frac{\partial^*}{\partial^* c^i}\left(\bar{L}_{v^j} \frac{\partial^* q^j}{\partial^* c^k}\right) - \frac{\partial^*}{\partial^* c^k}\left(\bar{L}_{v^j} \frac{\partial^* q^j}{\partial^* c^i}\right)$$

$$\equiv \frac{\partial^* \bar{L}_{v^j}}{\partial^* c^i} \frac{\partial^* q^j}{\partial^* c^k} - \frac{\partial^* \bar{L}_{v^j}}{\partial^* c^k} \frac{\partial^* q^j}{\partial^* c^i} = 0 \tag{6}$$

(The expression ω_{ik}, defined above, is clearly just the Lagrange bracket $[c^i, c^k]$.) We replace Equation (5) by the expression

$$\frac{\partial^* \bar{L}_{v^j}}{\partial^* t} \frac{\partial^* q^j}{\partial^* c^i} = \bar{L}_{q^j} \frac{\partial^* q^j}{\partial^* c^i} + \mu_i^{jk} \omega_{jk} \tag{7}$$

for some μ_i^{jk}. Then

$$\frac{\partial^* \bar{L}}{\partial^* c^i} = \bar{L}_{q^j} \frac{\partial^* q^j}{\partial^* c^i} + \bar{L}_{v^j} \frac{\partial^* v^j}{\partial^* c^i}$$

$$= \frac{\partial^* L_{v^j}}{\partial^* t} \frac{\partial^* q^j}{\partial^* c^i} + \bar{L}_{v^j} \frac{\partial^{*2} q^j}{\partial^* t \partial^* c^i} - \mu_i^{j\ell} \omega_{j\ell}. \tag{8}$$

so that

$$\frac{\partial^{*2} \bar{L}}{\partial^* c^i \partial^* c^k} = \frac{\partial^{*2}}{\partial^* c^k \partial^* t}\left(\bar{L}_{v^j} \frac{\partial^* q^j}{\partial^* c^i}\right) - \frac{\partial^*}{\partial^* c^k} (\mu_i^{j\ell} \omega_{j\ell}), \tag{9}$$

which implies that

$$\frac{\partial^{*2} \bar{L}}{\partial^* c^i \partial^* c^k} - \frac{\partial^*}{\partial^* t}\left(\bar{L}_{v^j} \frac{\partial^{*2} q^j}{\partial^* c^i \, {}^* c^k}\right) = \frac{\partial^*}{\partial^* t}\left(\frac{\partial^* \bar{L}_{v^j}}{\partial^* c^k} \frac{\partial^* q^j}{\partial \, c^i}\right) -$$

$$- \frac{\partial^*}{\partial^* c^k} (\mu_i^{j\ell} \omega_{j\ell}). \tag{10}$$

Since the left-handed side of Equation (10) is symmetric in i and k, we have

$$\frac{\partial^*}{\partial^* t}\omega_{ik} = \frac{\partial^*}{\partial^* c^i}(\mu_k^{j\ell}\omega_{j\ell}) - \frac{\partial^*}{\partial^* c^k}(\mu_i^{j\ell}\omega_{j\ell}). \tag{11}$$

If we choose $\mu_k^{j\ell} \equiv 0$, then we obtain the result that (5) implies:

$$\frac{\partial^*}{\partial^* t}\omega_{ik} = 0. \tag{12}$$

It should be noted that the operator $\partial^*/\partial^* t$ appearing in this section, being a differentiation along the curves $q = q(t)$ determined by (2), i.e., along the curves c^i=constant, i=1, ..., n, is distinct from the operator $\partial/\partial t$ used in the previous section, which was a differentiation along the curves q^i=constant, i=1, ..., n. If we choose

$$\mu_k^{j\ell} \equiv v^r \frac{\partial c^j}{\partial q^r}\delta_k^\ell,$$

where

$$\frac{\partial c^j}{\partial q^r}\frac{\partial^* q^r}{\partial^* c^i} = \delta_i^j ,$$

then

$$\frac{\partial^*}{\partial^* t}\omega_{ik} = \frac{\partial^*}{\partial^* c^i}\left(v^r \frac{\partial c^j}{\partial q^r}\right)\omega_{jk} - \frac{\partial^*}{\partial^* c^k}\left(v^r \frac{\partial c^j}{\partial q^r}\right)\omega_{ji} +$$

$$+ v^r \frac{\partial^* c^j}{\partial^* q^r}\left(\frac{\partial^*\omega_{jk}}{\partial^* c^i} - \frac{\partial^*\omega_{ji}}{\partial^* c^k}\right), \tag{13}$$

so that since

$$\frac{\partial^*\omega_{jk}}{\partial^* c^i} - \frac{\partial^*\omega_{ji}}{\partial^* c^k} = \frac{\partial^*\omega_{ik}}{\partial^* c^j} ,$$

we have

$$\frac{\partial^*}{\partial^* t}\,\omega_{ik} - v^r\,\frac{\partial c^j}{\partial q^r}\,\frac{\partial^*\omega_{ik}}{\partial^* c^j} = \frac{\partial^*}{\partial^* c^i}\left(v^r\,\frac{\partial c^j}{\partial q^r}\right)\omega_{jk} - $$

$$- \frac{\partial^*}{\partial^* c^k}\left(v^r\,\frac{\partial c^j}{\partial q^r}\right)\omega_{ji}. \qquad (14)$$

Note that the left-hand side of Equation (14) is just the operator $\partial/\partial t$, in the sense of Section 5, since

$$\frac{\partial^*}{\partial^* t} \equiv \frac{\partial}{\partial t} + v^r\frac{\partial}{\partial q^r} \equiv \frac{\partial}{\partial t} + v^r\,\frac{\partial c^j}{\partial q^r}\,\frac{\partial^*}{\partial^* c^j}\ . \qquad (15)$$

Thus Equation (14) is

$$\frac{\partial}{\partial t}\,\omega_{ik} = \frac{\partial^*}{\partial^* c^i}\left(v^r\frac{\partial c^j}{\partial q^r}\right)\omega_{jk} - \frac{\partial^*}{\partial^* c^k}\left(v^r\frac{\partial c^j}{\partial q^r}\right)\omega_{ji} \qquad (16)$$

and Equation (7) is

$$\frac{\partial^*\bar{L}_{v^k}}{\partial^* t} = \bar{L}_{q^k} + \frac{\partial^* q^r}{\partial^* t}\left(\frac{\partial c^j}{\partial q^r}\,\frac{\partial^*\bar{L}_{v^k}}{\partial^* c^j} - \frac{\partial c^j}{\partial q^r}\,\frac{\partial^*\bar{L}_{v^k}}{\partial^* c^j}\right), \qquad (17)$$

i.e.

$$\frac{\partial\bar{L}_{v^k}}{\partial t} = \bar{L}_{q^k} - v^r\frac{\partial L_{v^r}}{\partial q^k}\ , \qquad (18)$$

for this choice of μ_k^{ij}. Note that Equation (18) is identical to 5.3.

A summary of Sections 5 and 6 will be presented below for the convenience of the reader. We have defined two functions,

$$\theta_{ij} \equiv \frac{\partial}{\partial q^i}\,\bar{L}_{v^j} - \frac{\partial}{\partial q^j}\,\bar{L}_{v^i} \qquad (19)$$

and

$$\omega_{ik} \equiv \frac{\partial^*\bar{L}_{vj}}{\partial^* c^i}\,\frac{\partial^* q^j}{\partial^* c^k} - \frac{\partial^*\bar{L}_{vj}}{\partial^* c^k}\,\frac{\partial^* q^j}{\partial^* c^i}\ , \qquad (20)$$

and have found the behavior of these functions in two different sets of equations, each written both in the Eulerian and Lagrangian descriptions. The two sets of equations are equivalent if the corresponding initial conditions satisfy $\theta_{ij} = 0$ or $\omega_{ik} = 0$.

SUMMARY 1. The first set of equations is

$$\frac{\partial}{\partial t} \bar{L}_{v^i} = \bar{L}_{q^i} - v^j \frac{\partial \bar{L}_{v^j}}{\partial q^i} \tag{21}$$

in the Eulerian description, and

$$\frac{\partial^* \bar{L}_{v^k}}{\partial^* t} = \bar{L}_{q^k} + v^r \left(\frac{\partial c^j}{\partial q^r} \frac{\partial^* \bar{L}_{v^k}}{\partial^* c^j} - \frac{\partial c^j}{\partial q^k} \frac{\partial^* \bar{L}_{v^r}}{\partial c^j} \right) \tag{22}$$

in the Lagrangian description. For this set we have found

$$\frac{\partial}{\partial t} \theta_{ik} = 0 \tag{23}$$

and

$$\frac{\partial}{\partial t} \omega_{ik} = \frac{\partial^*}{\partial^* c^i} \left(\frac{\partial^* q^r}{\partial^* t} \frac{\partial c^j}{\partial q^r} \right) \omega_{jk} - \frac{\partial^*}{\partial^* c^k} \left(\frac{\partial^* q^r}{\partial^* t} \frac{\partial c^j}{\partial q^r} \right) \omega_{ji}. \tag{24}$$

SUMMARY 2. The second set of equations is

$$\frac{\partial}{\partial t} \bar{L}_{v^i} + v^j \frac{\partial \bar{L}_{v^i}}{\partial q^j} = \bar{L}_{q^i} \tag{25}$$

in the Eulerian description, and

$$\frac{\partial^*}{\partial^* t} \bar{L}_{v^j} = \bar{L}_{q^i} \tag{26}$$

in the Lagrangian description. For this set we have found

$$\frac{\partial^*}{\partial^* t} \theta_{ik} = \frac{\partial v^j}{\partial q^i} \theta_{kj} - \frac{\partial v^j}{\partial q^k} \theta_{ij} \tag{27}$$

and

$$\frac{\partial^*}{\partial^* t}\,\omega_{ik} = 0 \tag{28}$$

It is clear that other choices of λ_k^{ij} or μ_k^{ij} in both Sections 5 and 6 will lead to new systems different from those in Summaries (1) and (2), and equivalent to them if $\theta_{ij} = 0$, $\omega_{ij} = 0$ initially. These will not be investigated here.

7. THE HAMILTONIAN DESCRIPTION

By means of a change of variables it is possible to give a considerably clearer derivation of Equation (5.8). In terms of the Hamiltonian variables, Equation (6.1) becomes

$$d\phi = -H\,dt + p_i\,dq^i \tag{1}$$

hence cross-differentiation gives

$$\frac{\partial H}{\partial q^i} + \frac{\partial p_i}{\partial t} = 0 \tag{2}$$

and

$$\theta_{ik} \equiv \frac{\partial p_k}{\partial q^i} - \frac{\partial p_i}{\partial q^k} = 0 \tag{3}$$

It is clear that

$$\frac{\partial^2 p_k}{\partial t \partial q^i} = -\frac{\partial^2 H}{\partial q^i \partial q^k} = \frac{\partial^2 p_i}{\partial t \partial q^k}\,, \tag{4}$$

and hence

$$\frac{\partial}{\partial t}\,\theta_{ik} = 0\,, \tag{5}$$

which is just Equation (5.8).

8. CHARACTERISTICS

We recall the definition of the characteristics of a system of first-order linear partial differential equations

$$\sum_{i,\, j=1}^{\min} \alpha_k^{ij} \frac{\partial u_j}{\partial x^i} + \alpha_k = 0, \qquad k=1, \ldots, n \tag{1}$$

The characteristic elements in $(x^1, \ldots, x^m)$ are the annihilators of the complex-valued solutions $(\xi_1, \ldots, \xi_m)$ of the equation

$$\det \sum_{i=1}^{m} \xi_i \alpha_k^{ij} = 0 \ , \tag{2}$$

i.e., the solution $(\xi_1, \ldots, \xi_m)$ defines an element $(x^1, \ldots, x^m)$, which is its "orthogonal complement". At any point in the space of all the variables, if Equation (2) contains a factor $\Sigma_{i=1}^m v^i \xi_i$ which is linear in $(\xi_1, \ldots, \xi_m)$, the coefficient v^i appearing in this factor may be taken as the ith direction number of a characteristic direction in $(x^1, \ldots, x^m)$ at that point. For m=2, the fundamental theorem of algebra ensures that n linear factors always exist, and so the preceding definitions coincide. In this case there exist linear combinations of the members of Equation (1) in which every unknown is differentiated only in one given characteristic direction. For $m > 2$, there exist linear combinations of the members of Equation (1) in which every unknown is differentiated in the same direction only if a characteristic direction exists, this being the direction of differentiation, but not conversely. (A necessary and sufficient condition can easily be given for the existence of a linear combination of the members of Equation (1) in which every unknown is differentiated in the same direction, but it involves the existence of simultaneous solutions of (m-1)(n-1) quadratic equations in n-1 unknowns, and so it is not very useful.)

If a characteristic direction $(v^1, \ldots, v^m)$ exists and varies continuously in the neighborhood of some point in the space of all the variables, then the curves defined by

$$\frac{dx^1}{dv^1} = \cdots = \frac{dx^m}{dv^m} \tag{3}$$

are called the characteristic curves in $(x^1, \ldots, x^m)$.

We return to the Eulerian equations of Section 5, which may be written as

$$\frac{\partial}{\partial t} \bar{L}_{v^i} = \bar{L}_{q^i} + (\lambda-1)v^j \frac{\partial \bar{L}_{v^j}}{\partial q^i} - \lambda v^j \frac{\partial \bar{L}_{v^i}}{\partial q^j} \tag{4}$$

choosing $\lambda_{i}^{jk} = \lambda\delta_{i}^{k}v^j$; i.e.,

$$\delta_{i}^{j} \frac{\partial p_j}{\partial t} + [(1-\lambda)\delta_{i}^{k}v^j + \lambda\delta_{i}^{j}v^k] \frac{\partial p_j}{\partial q^k} - r_i = 0 \tag{5}$$

where $p_i = \bar{L}_{v^i}$ and $r_i = \bar{L}_{q^i}$. If one sets $t = x^{n+1}$, and imagines that an index i=n+1 is attached to the first term of Equation (5), then Equation (5) has the form of Equation (1). To find $(\xi, \xi_1, \ldots, \xi_n)$, given its characteristic elements in $(t, x^1, \ldots, x^n)$, set

$$\begin{aligned}
&\det[\xi\delta_{i}^{j} + (1-\lambda)\delta_{i}^{k}\xi_k v^j + \lambda\delta_{i}^{j}\xi_k v^k] \\
&\quad \equiv \det[(\xi+\lambda\xi_k v^k)\delta_{i}^{j} + (1-\lambda)\xi_i v^j] \\
&\quad \equiv (\xi+\lambda\xi_k v^k)^n + (\xi+\lambda\xi_k v^k)^{n-1}(1-\lambda)\xi_k v^k \\
&\quad \equiv (\xi+\lambda\xi_k v^k)^{n-1}(\xi+\xi_k v^k) = 0
\end{aligned} \tag{6}$$

Thus Equation (5) actually possesses n characteristic directions in $(x^1, \ldots, x^n)$, n-1 of them coinciding with $(1, \lambda v^1, \ldots, \lambda v^n)$, the remaining one being $(1, v^1, \ldots, v^n)$ The characteristic curve corresponding to the direction $(1, v^1, \ldots, v^n)$ is determined by

$$\frac{dq^i}{dt} = v^i(q, t), \quad i=1, \ldots, n, \tag{7}$$

which is the particle flow.

There are other choices of λ_i^{jk} for which the Eulerian equations contain the characteristic curve (7). For instance, if $\lambda_i^{jk} = d_i^k v^j$, where d_i^k is a diagonal matrix with $d_i^i = 1$ for some i (no summation), then the resulting equations contain the characteristic curve (7), although, in general, no other characteristic directions exist at all. However, we shall later specialize (5), instead of generalizing it, by taking $\lambda=1$; in this case there are n characteristic curves, all identical with (7).

One usually defines a characteristic direction in $(x^1, \ldots, x^m)$ for a given system, Equation (1), in the hope of finding a linear combination of the equations of that system in which the unknowns are differentiated only in the given characteristic direction. In the case $\lambda=1$, Equation (5) becomes Equation (5.9) which we repeat here for convenience:

$$\frac{\partial p_i}{\partial t} + v^k \frac{\partial p_i}{\partial q^k} - r_i = 0 \tag{8}$$

It is immediately clear that every differentiation occurs in the unique characteristic direction. In general, given a characteristic direction $(v^1, \ldots, v^m)$, one may find the desired linear combination, if it exists, by solving

$$\sum_{k=1}^{m} \xi_k \alpha_i^k \eta^i = 0 , \tag{9}$$

for $\eta^1, \ldots, \eta^n$, where $\Sigma_{i=1}^m \xi_i v^i = 0$, which gives the combination

$$\sum_{k=1}^{m} \eta^i \alpha_i^{kj} \frac{\partial u_j}{\partial x^k} + \eta^i \alpha_i = 0. \tag{10}$$

If this combination is independent of $(\xi_1, \ldots, \xi_m)$, then every unknown appearing in it is indeed differentiated in the characteristic direction $(v^1, \ldots, v^m)$. This is seldom the case. Returning to (8), we see that for the characteristic direction $(1, v^1, \ldots, v^n)$, this method gives

$$\{\xi\delta_i^j + \xi_k[(1-\lambda)\delta_i^k v^j + \lambda\delta_i^j v^k]\}\eta^i$$
$$\equiv (1-\lambda)(\xi_i v^j - \xi_k v^k \delta_i^j)\eta^i = 0, \tag{11}$$

taking account of the definition, $\xi + \xi_k v^k = 0$. Hence for $\lambda \neq 1$, η^i is proportional to v^i. Thus, one may take the linear combination

$$v^j \frac{\partial p_j}{\partial t} + [(1-\lambda)v^j v^k + \lambda v^j v^k] \frac{\partial p_j}{\partial q^k} - v^i r_i$$
$$\equiv v^j \frac{\partial p_j}{\partial t} + v^j v^k \frac{\partial p_j}{\partial q^k} - v^i r_i = 0 \tag{12}$$

and note that every differentiation is indeed in the direction $(1, v^1, \ldots, v^n)$.

There are no similar results corresponding to the characteristic directions $(1, \lambda v^1, \ldots, \lambda v^n)$, for $\lambda \neq 1$.

We return to the case $\lambda=1$ and (8). It has already been noted that this system has the form of Jacobi's simultaneous equations; now we make use of the fact that such a system can be treated essentially like one first-order partial differential equation in a single unknown.

We attempt to find a relation connecting dp_i, dq^i and dt which is independent of the partial differential coefficients $\partial p_i/\partial q^k$ and $\partial p_i/\partial t$. To this end, note that (7) implies that

$$dp_i = \frac{\partial p_i}{\partial q^k} dq^k + \frac{\partial p_i}{\partial t} dt = \frac{\partial p_i}{\partial q^k} dq^k + \left(r_i - v^k \frac{\partial p_i}{\partial q^k}\right) dt \tag{13}$$

i.e.

$$(dp_i - r_i dt) = \frac{\partial p_i}{\partial q^k} (dq^k - v^k dt). \tag{14}$$

Since det $\partial p_i/\partial q^k \neq 0$ by assumption, the only solutions of (14) which are independent of $\partial p_i/\partial q^k$, and hence of $\partial p_i/\partial t$, are given by

$$dq^k - v^k dt = 0,$$
$$dp_i - r_i dt = 0. \tag{15}$$

The first set of these equations defines the characteristics in $(t, q^1, \ldots, q^n)$ as before. Eliminating $v^1, \ldots, v^n$ from both sets of equations, it is clear that (14) gives rise to the Euler-Lagrange variational equations. By analogy with the definition for one first-order partial differential equation, there is some justification for calling the solutions of (14) the characteristics of (8) in the space of all the variables.

It will be worth while to restate the results of this section. First, we have examined a one-parameter family of systems of equations, Equation (4), obtained by adding a linear combination of terms θ_{ij}, defined in Equations (5.1) through (5.3). For every value of the parameter, the system (4) possesses n characteristic directions, one of them given by (7), which is the same as the first members of (14); and there exists a linear combination of the members of (4) in which every unknown is differentiated in this characteristic direction.

For one particular value of the parameter, namely, the value $\lambda=1$, giving rise to (5.9), which has been rewritten as (8), all n characteristic directions are identical and are defined by (5). In this case every solution of the system satisfies the relation $dp_i - r_i dt = 0$ along the characteristics. Conversely, the charateristics and the relations $dp_i - r_i dt = 0$ are those curves in the space of all the variables which generate the solution surfaces, i.e., which are independent of the partial differential coefficients. More specifically, suppose that a point $(t, q^1, \ldots, q^n, p_1, \ldots, p_n)$ on one of the (n+1)-dimensional integral manifolds of Equation (8) is specified. Then the curve through this point, which is defined by (14), lies entirely on the given integral manifold, and, indeed, on any integral manifold in the (2n+1)-dimensional space containing the point $(t, q^1, \ldots, q^n, p_1, \ldots, p_n)$. Given any n-dimensional hypersurface in the space, which satisfies (14) nowhere, an integral manifold containing this hypersurface is simply the locus of points lying on those solutions of (14) which meet the hypersurface.

Finally, the projections of (14) on $(t, q^1, \ldots, q^n)$ are just the Euler-Lagrange variational equations. Since the variational equations can also be obtained merely by writing Equation (8) in the Lagrangian form, Equation (8) shows that particles of the fluid whose motion it describes flow along the characteristic curves.

BIBLIOGRAPHY AND COMMENTS

Section 1. These results were obtained by Howard Osborn in 1953 and never published.

Chapter V

QUASILINEARIZATION AND A NEW METHOD OF SUCCESSIVE APPROXIMATIONS

1. INTRODUCTION

The purpose of this chapter is to present a new method for treating a class of partial differential equations. The method depends upon the fact that the calculus of variations can be considered to be a continuous multistage decision process. Hence, we use the theory of dynamic programming to obtain a crucial relation.

We shall illustrate the method by considering the equation

$$\frac{\partial f}{\partial t} = c^2\left(\frac{\partial f}{\partial c} \right)^2/4 + f^2, \quad f(c, 0) = 0. \tag{1}$$

This equation is taken over the domain $c_1 \geq c \geq 0$, $T \geq t \geq 0$.

In Section 2, we present the fundamental relation upon which the method depends. In Section 3, we present the method of successive approximations we shall employ and obtain a bound for each approximant. In Section 4, we show the convergence of the approximants.

The results can be extended in various directions. Here we are only interested in showing the method.

We are using the basic idea of the theory of quasi-linearization.

2. THE FUNDAMENTAL VARIATIONAL RELATION

We shall use the following result, if

$$f = \min_u \left(\int_t^T q(u, s) + u^2 + u'^2 \right) ds \tag{1}$$

then

$$\frac{\partial f}{\partial t} = g(c, t) - (\frac{\partial r}{\partial c})^2/4 \qquad (2)$$

under suitable conditions in g.

Here the minimization is over $u(t) = c$.

3. SUCCESSIVE APPROXIMATIONS

We shall employ the method of successive approximations. Let

$$f_0 = \min_u \int_t^T (u^2+u^{\cdot 2})dt$$
$$f_{n+1} = \min_u \int_t^T (f_n^2 + u^2 + u^{\cdot 2})dt. \qquad (1)$$

Let us prove inductively that we have the bound $|f_n| \leq 2c^2$ for c and T small.

The relation for n=0 is easily obtained. For n equal to zero we can shift the limits so that the lower limit is zero. We then use the simple trial function ce^{-t}.

It is easily seen that if T is small then u is small. Hence, we can use the bound for f_n provided that c is small.

Once we have used this bound, we can again shift the limits of integration and use the same simple trial function. We see then that the bound is maintained.

4. CONVERGENCE

Let us now prove convergence.

We shall employ the following simple result which we have used repeatedly in discussing the functional equations of dynamic programming. Let

$$J_1 = \min_u \int_t^T (g_1(u, t) + u^{\cdot 2})dt$$
$$J_2 = \min_u \int_t^T (g_2(u, t) + u^{\cdot 2})dt. \qquad (1)$$

Let u_1 be a function which furnishes the minimum value of J_1. Let u_2 be a function which furnishes the minimum value of J_2. Then we have

$$\begin{aligned} J_1(u_1) &\leq J_2(u_1) \\ J_2(u_2) &\leq J_1(u_2). \end{aligned} \tag{2}$$

Combining these relations we have

$$|J_1 - J_2| \leq T \max |g_1 - g_2|. \tag{3}$$

Using this relation, and the fact that c and T are small, we have

$$|f_{n+1} - f_n| \leq \lambda |f_n - f_{n-1}|, \tag{4}$$

where λ is less than one.

Hence we have uniform convergence of the sequence f_n.

Hence, the limit function f satisfies the equation

$$f = \min_u \int_t^T (f^2 + u^2 + \dot{u}^2)dt. \tag{5}$$

The same simple dynamic programming argument used above shows that f satisfies the nonlinear partial differential equation. A simple change of the t variable is required to get the equation in the form given in the introduction.

BIBLIOGRAPHY AND COMMENTS

Section 1. These results were given in:

R. Bellman, 'A New Method for Treating a Class of Nonlinear Partial Differential Equations', *Nonlinear Analysis, Theory, Methods, and Applications*, Vol. 3, No. 5, pp.721-722.

For the theory of quasilinearization, see:

R. Bellman, *Methods of Nonlinear Analysis, Vol. II*, Academic Press, Inc., New York, 1973.

Practical results using quasilinearization are given in:

E.S. Lee, *Quasilinearization and Invariant Imbedding*, Academic Press, Inc., New York, 1968.

Chapter VI

THE VARIATION OF CHARACTERISTIC VALUES AND FUNCTIONS

1. INTRODUCTION

The Green's function associated with the second order equation

$$u'' + q(x)u = 0, \quad u(a) = u(1) = 0, \tag{1}$$

has been discussed. Analogous methods are used in the following chapter to obtain the Hadamard variational formula.

In this chapter, we wish to present some extensions of these results. Introducing the parameter λ, we consider the general equations

$$(p(x)u')' + (r(x) + \lambda q(x))u = 0, \quad u(a)=0, \quad u(1)=0, \tag{2}$$

obtaining, variational equations for the resolvent operator as a function of a. Utilizing the meromorphic nature of the operator as a function of λ, we are able in this way to derive variational equations for the characteristic values and functions.

Corresponding results are obtained for the vector-matrix system

$$(p(t)x')' + \lambda Q(t)x = 0. \tag{3}$$

In both cases, certain assumptions have to be made concerning the coefficient functions, $q(x)$, and $Q(t)$, in order to be able to consider the differential equations as the Euler-Lagrange equations of associated variational problems. Since, however, we know, from the results of Miller and Schiffer, concerning Green's functions for general linear differential operators of order n (where quite different methods are employed) and from corresponding results for Fredholm operators and Jacobi matrices, that the relations obtained hold under far weaker assumptions, the interesting problem arises of deriving the more general results by variational techniques. In this chapter, we present a method of analytic continuation which reduces the case of continuous

$q(x)$ to that of negative continuous $q(x)$, and the case of continuous symmetric $Q(t)$ to negative definite $Q(t)$.

In a brief section, we indicate how similiar methods may be applied to obtain corresponding results for Fredholm kernels and the associated characteristic values and functions.

2. VARIATIONAL PROBLEM

Consider the boundary value problem

$$(p(x)u')'+q(x)u = v(x), \quad a < x < 1, \quad u(a)=0,$$
$$u(1) + \alpha u'(1) = 0. \tag{1}$$

We shall assume that p, q, and v are continuous functions on the closed interval [a, 1]. If the corresponding Sturm-Liouville problem with q replaced by λq and v by 0 does not have 1 as a characteristic value, then the unique solution of (1) can be represented in the form

$$u(x) = \int_a^1 K(x, y, a)v(y)dy. \tag{2}$$

The function $K(x, y, a)$ is called the Green's function for the boundary value problem (1). We wish to study the dependence of K upon a by means of the functional equation method of dynamic programming. We imbed (1) in the system

$$(p(x)u')' + q(x)u = v(x), \quad a < x < 1, \quad u(a) = c,$$
$$u(1) + \alpha u'(1) = 0. \tag{3}$$

A solution of this system can be expressed in the form

$$u(x) = u_0(x) + c\phi(x) \tag{4}$$

where $u_0(x)$ is a solution of (1) and hence equal to the right side of (2) while $\phi(x)$ is a solution of the system

$$(p(x)\phi')' + q(x)\phi = 0, \ \phi(a) = 1,$$
$$\phi(1) + \alpha\phi'(1) = 0 \tag{5}$$

We also consider a variational problem associated with the system (3), the problem of maximizing $J(u, v)$ over all u for which $u(a) = c$ where

$$J(u, v) = \int_a^1 (q(x)u^2 - p(x)u'^2 - 2uv(x))dx - \frac{p(1)}{\alpha}(u(1))^2 \tag{6}$$

This variational problem has the property that a u yielding the maximum is a solution of the system (3). To prove this, we replace u by $u + \varepsilon\eta$ in (6), where $\eta(x)$ is a function such that $\eta(a) = 0$. We obtain

$$J(u+\varepsilon\eta, v) = J(u, v) + 2\varepsilon\left\{\int_a^1 (qu\eta - pu'\eta' - v\eta)dx - \frac{p(1)}{\alpha}u(1)\eta(1)\right\} +$$
$$+ \varepsilon^2\left[\int_a^1 (q\eta^2 - p\eta^2)dx - \frac{p(1)}{\alpha}\eta(1)^2\right].$$

In order to have u yield the maximum, the ε term must vanish for all functions η for which $\eta(a) = 0$. Hence

$$\int_a^1 (qu+(pu')' - v\eta(x)dx - \frac{p(1)}{\alpha}\eta(1)(u(1)+\alpha u'(1)) = 0.$$

Consequently, a u yielding the maximum must satisfy the boundary condition at 1 and must satisfy the differential equation in the interior of the interval $[a, 1]$. Hence, a solution of the variational problem provides a solution of the boundary value problem (3).

In order to use the variational approach one must make an assumption on p and q sufficient to guarantee the existence of a maximum. It is sufficient, for example, to assume that $p(x)$ is positive on the closed interval $[a, 1]$ and that the smallest characteristic value of the Sturm-Liouville problem

$$(pu') + \lambda qu = 0, \quad u(a) = 0, \quad u(1) + \alpha u'(1) = 0$$

is larger than 1. If $q(x)$ is uniformly positive, $d \geq q(x) > 0$, over $0 \le x \le a$, this condition holds if a is sufficiently small, or if d is sufficiently large. This assumption guarantees the existence of a unique maximum of $J(u, v)$. In §7, we shall show by analytic continuation how the results we prove can be freed of this restrictive hypothesis.

3. DYNAMIC PROGRAMMING APPROACH

We now let

$$f(a, c) = \max_{u(a)=c} J(u, v)$$

and derive a partial differential equation for f by means of a technique of dynamic programming.

We regard u as describing a policy. The variable c describes the state of the system at a. The result of following the policy u for a time interval $[a, a+\Delta]$ is to transform c into a new initial state $u(a+\Delta)$ for the interval $[a+\Delta, 1]$. Translated into a formula, the principle of optimality yields the equation

$$f(a, c) = \max_{u(a)=c} \left\{ f((a+\Delta), u(a+\Delta)) + \int_a^{a+\Delta} (q(x)u(x)^2 - p(x)u'(x)^2 - 2u(x)v(x))dx \right\}.$$

We proceed formally, assuming that $f(a, c)$ has continuous partial derivatives and that the maximizing u has a continuous derivative. Then as $\Delta \to 0$,

$$f(a+\Delta, u(a+\Delta)) = f(a, c) + \frac{\partial f}{\partial a}(a, c)\Delta + \frac{\partial f}{\partial c}(a, c)u'(a)\Delta + C(\Delta).$$

Consequently, as $\Delta \to 0$, we have

$$-\frac{\partial f}{\partial a}(a, c) = \max_{u(a)=c} \left\{ \frac{\partial f}{\partial c}(a, c)u'(a) + q(a)c^2 - p(a)u'(a)^2 - 2cv(a) - O(1) \right\}.$$

The quadratic

$$\frac{\partial f}{\partial c} u' - p(a)u'^2$$

takes its maximum value at

$$u' = \frac{1}{2p(a)} \frac{\partial f}{\partial c}(a, c).$$

Hence we obtain the partial differential equation

$$-\frac{\partial f}{\partial a}(a, c) = \frac{1}{4p(a)}\left(\frac{\partial f}{\partial c}(a, c)\right)^2 - 2cv(a) + c^2 q(a). \qquad (1)$$

4. VARIATION OF THE GREEN'S FUNCTION

Let u be the function which maximizes J(u, v) for given a and c. By using (2.2) and (2.4), we can find an equation which connects f(a, c) with the Green's function K(x, y, a). We have

$$f(a, c) = \int_a^1 (qu^2 - pu'^2 - 2uv)dx - \frac{p(1)}{\alpha}(u(1))^2$$

$$= \int_a^1 u(qu + (pu')' - v)dx - \int_a^1 uv\, dx - u(1)p(1)u'(1) +$$

$$+ u(a)p(a)u'(a) - \frac{p(1)}{\alpha}(u(1))^2 \qquad (1)$$

$$= -\int_a^1 uv\, dx + cp(a)u'(a) - \frac{p(1)}{\alpha}u(1)(u(1) + \alpha u'(1))$$

$$= -\int_a^1 u_0 v\, dx - c\int_a^1 \phi v\, dx + cp(a)u_0'(a) + c^2 p(a)\phi'(a).$$

using the fact that u_0 satisfies (2.1) and ϕ satisfies (2.5), we obtain by integration by parts

$$\int_a^1 \phi v\,dx = \int_a^1 \phi((pu_0')' + qu_0)dx$$

$$= \int_a^1 u_0((p\phi')'+q\phi)dx+\phi(1)p(1)u_0'(1) - \phi(a)p(a)u_0'(a) - \phi'(1)p(1)u_0(1) + \phi'(a)p(a)u_0(a).$$

Since $u_0(a) = 0$, $\phi(a) = 1$, and u and ϕ satisfy the same homogeneous boundary condition at 1, we conclude that

$$\int_a^1 \phi v\,dx = -p(a)u_0'(a)+(\phi(1)p(1)u_0'(1) - \phi'(1)p(1)u_0(1)) = -p(a)u_0'(a). \tag{2}$$

Thus, the second and third term on the last line of (1) are equal. It turns out to be more convenient to use the integral expression. We have

$$f(a, c) = -\int_a^1 u_0 v\,dx - 2c\int_a^1 \phi v\,dx + c^2 p(a)\phi'(a)$$

$$= -\int_a^1\int_a^1 K(x, y)v(x)v(y)\,dx\,dy - 2c\int_a^1 \phi v\,dx + c^2 p(a)\phi'(a). \tag{3}$$

We observe that we can express ϕ in terms of a partial derivative of K. By (2), we have

$$\int_a^1 \phi(y)v(y)\,dy = -p(a)u_0'(a) = -p(a)\int_a^1 \frac{\partial K}{\partial x}(a, y, a)v(y)\,dy$$

for all continuous functions v. Hence

$$\phi(y) = -p(a)\,\frac{\partial K}{\partial x}(a, y, a). \tag{4}$$

Combining the expression for f(a, c) given by (3) with the partial differential equation (3.1), we obtain upon equating terms independent of c,

$$\int_a^1 \int_a^1 \frac{\partial K}{\partial a}(x, y, a)v(x)v(y)\,dx\,dy = \frac{4}{4p(a)}\left(\int_a^1 \phi(x)v(x)\,dx\right)^2$$

$$= \frac{1}{p(a)}\int_a^1 \int_a^1 \phi(x)\phi(y)v(x)v(y)\,dx\,dy.$$

Now if we "equate coefficients" of v(x)v(y) we obtain the relation

$$\frac{\partial K}{\partial a}(x, y, a) = \frac{1}{p(a)}\,\phi(x)\phi(y)$$

$$= p(a)\,\frac{\partial K}{\partial y}(x, a, a)\,\frac{\partial K}{\partial x}(a, y, a). \tag{5}$$

This formal equating of coefficients can be justified with the use of the symmetry of the Green's function and an assumption such as continuity of

$$\frac{\partial K}{\partial a}(x, y, a) - \frac{1}{p(a)}\,\phi(x)\phi(y).$$

We shall prove this in § 5.

In the consideration of the boundary value problem (2.3), we have not considered the boundary conditions u(1) = 0 and u'(1) = 0. The condition u(1) = 0 corresponds formally to $\alpha = 0$; the condition u'(1) = 0 corresponds formally to $\alpha = \infty$ in the condition $u(1) + \alpha u'(1) = 0$. If in the variational problem of maximizing J(u, v) we omit the term

$$\frac{-p(1)}{\alpha}\,u(1)^2$$

and add the constraint u(1) = 0 or u'(1) = 0, respectively,

the development proceeds as before, except all terms which involve α vanish. The final results obtained are the same.

5. JUSTIFICATION OF EQUATING COEFFICIENTS

The formal procedure used in the derivation of (4.5) can be justified by the following lemma.

LEMMA. Let $F(x, y)$ be a continuous function on the region $a < x < 1$, $a < y < 1$ and suppose that $F(x, y) = F(y, x)$. Then if

$$\int_a^1 \int_a^1 F(x, y)v(x)v(y)\, dx\, dy = 0$$

for all continuous functions v, the function $F(x, y) \equiv 0$.

Proof. First, we show that F vanishes on the diagonal, i.e., $F(\zeta, \zeta) = 0$. To do this, we take a sequence of continuous functions v, with ε positive and tending to 0, each of which vanishes identically outside the interval $[\zeta - \varepsilon, \zeta + \varepsilon]$ and integrates to 1 over the portion of $[\zeta - \varepsilon, \zeta + \varepsilon]$ which is in the interval $[a, 1]$. Passing to the limit, we find that $F(\zeta, \zeta) = 0$.

To prove that $F(\zeta, \eta) = 0$ when $\zeta \neq \eta$ we use functions v, with $\varepsilon < |\zeta - \eta|/2$ which integrate to 1 over the portion of $[\zeta - \varepsilon, \zeta + \varepsilon]$ in the interval $[a, 1]$ and also integrate to 1 over the portion of $[\eta - \varepsilon, \eta + \varepsilon]$ in $[a, 1]$ but which vanish identically outside these two intervals. Letting $\varepsilon \to 0$, we find

$$F(\zeta, \eta) + F(\eta, \zeta) + F(\zeta, \zeta) + F(\eta, \eta) = 0.$$

since

$$F(\zeta, \zeta) = F(\eta, \eta) = 0 \text{ and } F(\zeta, \eta) = F(\eta, \zeta),$$

we have

$$F(\zeta, \eta) = 0.$$

6. CHANGE OF VARIABLE

In this section, we shall make a change of variable which leads to a different expression for $\phi(x)$ and hence all alternate expression for the variation of the Green's function. We make the change of variable.

$$u = c\frac{(1 + \alpha + x)}{(1 + \alpha - a)} + w$$

so that $w(a) = 0$ and $w(1) + \alpha w'(1) = 0$ when u satisfies the boundary conditions $u(a) = c$, $u(1) + \alpha u'(1) = 0$. (This transformation is valid also in case $\alpha = 0$; for the condition $u'(1) = 0$, i.e., $\alpha = \infty$, we set $u = c + w$.)

Then by (2.6),

$$J(u, v) = J(w, v) + 2c\int_a^1 \left(\frac{1 + \alpha - x}{1 + \alpha - a}\right) q(x)w(x)\, dx +$$

$$+ 2c\int_a^1 \frac{p(x)w'(x)}{1 + \alpha - a}\, dx - 2c\int_a^1 v(x)\left(\frac{1 + \alpha - x}{1 + \alpha - z}\right)dx - 2c\,\frac{p(1)w(1)}{1 + \alpha - a} +$$

$$+ c^2\left\{\int_a^1 \left(\frac{1 + \alpha - x}{1 + \alpha - a}\right)^2 q(x)\, dx - \int_a^1 \frac{p(x)}{(1 + \alpha - a)^2}\, dx - \frac{p(1)\alpha}{(1 + \alpha - a)^2}\right\}.$$

Transforming the second integral by integration by parts, we obtain

$$J(u, v) = J\left(w,\ v - c\left[\left(\frac{1 + \alpha - x}{1 + \alpha - a}\right)q(x) - \frac{p'(x)}{1 + \alpha - a}\right]\right) + F(a, c, v)$$

where F is independent of w and u. Hence

$$f(a, c) = \max_{u(a)=c} J(u, v)$$

$$= \max_{w(a)=0} J\left(w,\ v - c\left[\left(\frac{1 + \alpha - x}{1 + \alpha - a}\right)q(x) - p'(x)\right]\right) + F(a, c, v).$$

The maximizing w is given by

$$w = \int_a^1 K(x, y, a)\left\{v(y) - c\left[\left(\frac{1 + \alpha - y}{1 + \alpha - a}\right)q(y) - \frac{p'(y)}{1 + \alpha - a}\right]\right\} dy.$$

Hence

$$u = \int_a^1 K(x, y, a)v(y)dy +$$

$$+c\left\{ \frac{1 + \alpha - x}{1 + \alpha - a} - \int_a^1 K(x, y, a\left[\left(\frac{1 + \alpha - y}{1 + \alpha - a}\right)q(y) - \frac{p'(y)}{1 + \alpha - a}\right]dy\right\}.$$

Thus by (2.4),

$$\phi(x) = \frac{1 + \alpha - x}{1 + \alpha - a} - \int_a^1 K(x, y, a)\left[\left(\frac{1 + \alpha - y}{1 + \alpha - a}\right)q(y) - \frac{p'(y)}{1 + \alpha - a}\right]dy.$$

Substituting this expression into (4.5), we find another formula for the variation of the Green's function,

$$\frac{\partial K(x, y, a)}{\partial a} = \frac{1}{p(a)}$$

$$\cdot\left\{\frac{1 + \alpha - x}{1 + \alpha - a} - \int_a^1 K(x, y, a)\left[\left(\frac{1 + \alpha - y}{1 + \alpha - a}\right)q(y) - \frac{p'(y)}{1 + \alpha - a}\right]dy\right\} \tag{1}$$

$$\cdot\left\{\frac{1 + \alpha - y}{1 + \alpha - a} - \int_a^1 K(x, y, a)\left[\left(\frac{1 + \alpha - x}{1 + \alpha - a}\right)q(x) - \frac{p'(x)}{1 + \alpha - a}\right]dx\right\}.$$

The cases $u(1) = 0$ and $u'(1) = 0$ are easily handled. The results are what would be obtained by setting $\alpha = 0$ or letting $\alpha \to \infty$ respectively.

7. ANALYTIC CONTINUATION

In order to extend the foregoing approach to the general case, we shall employ the method of analytic continuation.

Consider the equation

$$(pu')' + (t+q(x))u = v, \quad u(a) = c, \quad u(1)+\alpha u'(1) = 0, \tag{1}$$

where z is a positive quantity chosen large enough so that $-t+q(x) \leq 0$ in [a, 1].

Furthermore, we know from the Sturmian comparison theorems that z can be chosen large enough so that $\lambda = 1$ is not a characteristic value of the Sturm-Liouville problem

$$(pu')'+\lambda(t+q(x))u=0, \quad u(a) = 0, \; u(1)+\alpha u'(1)=0 \tag{2}$$

It follows that the inhomogeneous equation

$$(pu')'+(t+q(x))u = v, \quad u(a)=0, \; u(1)+\alpha u'(1)=0, \tag{3}$$

will then have a unique solution.

The Green's function associated with this problem will be a function of z. If we can show that this function is a meromorphic function of z, it will follow that the relations originally obtained under the assumption that z is sufficiently large will actually be valid for all z distinct from a set of poles.

In particular, the relations will be valid for $z = 0$, provided that $\lambda = 1$ is not a characteristic value of the equation

$$(pu')'+\lambda q(x)u = v, \quad u(a)=c, \; u(1)+\alpha u'(1)=0. \tag{4}$$

Generally speaking, the relations will be valid whenever they make sense.

8. ANALYTIC CHARACTER OF GREEN'S FUNCTION

Although it follows from well-known results in the theory of linear differential equations that the Green's function is a meromorphic function of z, we shall outline a proof here for the sake of completeness.

To simplify the notation let us write $\sigma(u) = 0$ for the boundary condition $u(1) + \alpha u'(1) = 0$. Furthermore we observe that the following considerations also cover the boundary condition $u'(1) = 0$, so that we could also take $\sigma(u) = u'(1)$.

To begin with, the two solutions of

$$(pu')' + (t+q(x))u = 0, \tag{1}$$

determined by the initial conditions

$$u_1(a) = 1, \quad u_1'(a) = 0, \quad u_2(a) = 0, \quad u_2'(a) = 1, \tag{2}$$

are entire functions of z for x in [0, 1]. Also for fixed x, their derivatives u_1' and u_2' are entire functions of z. Hence, in particular, $\sigma(u_2)$ is an entire function of z which does not vanish identically.

Concerning p(x) and q(x), we are assuming, as before, that $p(x) \geq d_1 > 0$ in [0, 1], that it has an integrable derivative, and that q(x) is integrable in [a, 1], and uniformly bounded.

The general solution of

$$(pu')' + (t+q(x))u = v \tag{3}$$

can be written

$$u = c_1u_1(x)+c_2u_2(x)- \frac{1}{p(a)} \int_a^x [u_1(x)u_2(y)-u_1(y)u_2(x)]v(y)dy \tag{4}$$

Imposing the boundary conditions

$$u(a) = 0, \quad \sigma(u) = 0,$$

we see that this particular solution has the form

$$u = \int_a^1 K(x, y)v(y)\, dy$$

where K(x, y), the desired Green's function, is given by

$$K(x, y) = \frac{1}{p(a)}\left[\frac{\sigma(u_1)u_2(y) - u_1(y)\sigma(u_2)}{\sigma(u_2)}\right]u_2(x), \qquad a < y < x,$$

$$= \frac{1}{p(a)}\left[\frac{\sigma(u_1)u_2(x) - u_1(x)\sigma(u_2)}{\sigma(u_2)}\right]u_2(y), \qquad a < x < y.$$

We see that for any x and y in [a, 1], K(x, y) and its partial derivatives with respect to x, y and a are meromorphic functions of z. Also, we observe that this formula provides a proof for the symmetry of the Green's function, a property which we have used in the previous sections.

9. ALTERNATE DERIVATION OF EXPRESSION FOR $\phi(x)$

The explicit representation for K(x, y) given above shows that

$$\left.\frac{\partial K}{\partial x}\right|_{x=a} = \frac{1}{p(a)}\left[\frac{\sigma(u_1)u_2(y) - u_1(y)\sigma(u_2)}{\sigma(u_2)}\right]$$

It follows that $-p(a)\,\partial K/\partial x(a, y, a)$ is the solution of the two-point boundary problem

$$(pu')' + qu = 0, \qquad u(a) = 1, \quad \sigma(u) = 0.$$

Hence we have

$$\phi(y) = -p(a)\,\frac{\partial K}{\partial x}\,(a,y,a).$$

The extension of the relation in (1) is the keystone of the treatment of Miller and Schiffer and the analogue of this relation for partial differential operators is essential in the derivation of the Hadamard variational formula given in Chapter 7.

10. VARIATION OF CHARACTERISTIC VALUES AND CHARACTERISTIC FUNCTIONS

Consider the Sturm-Liouville problem

$$(pu')'+(q(x)+\lambda r(x))u=v(x), \qquad a < x < 1, \quad \begin{array}{l} u(a)=0 \\ u(1)+\alpha u'(1)=0 \end{array} \tag{1}$$

This problem has a unique solution, when λ is not a characteristic value, which is given by

$$u(x) = \int_a^1 R(x,y,\lambda,a)v(y)dy.$$

The function R, called the resolvent, is a meromorphic function of λ with poles at the characteristic values λ_1, λ_2, λ_3, ... If $\psi_k(x)$ is the characteristic function associated with the characteristic value λ_k, then R has the representation

$$R(x,y,\lambda,a) = \sum_{k=1}^{\infty} \frac{\psi_k(x)\psi_k(y)}{\lambda-\lambda_k} \tag{2}$$

The problem (1) is obtained from (2.1) by replacing $q(x)$ by $q(x) + \lambda r(x)$. From (4.5) we obtain the equation

$$\frac{R(x,\ y,\ \lambda,\ a)}{\partial a} = p(a)\frac{\partial R}{\partial x}(a,\ y,\ \lambda,\ a)\frac{\partial R}{\partial y}(x,\ a,\ \lambda,\ a). \tag{3}$$

This equation is valid for all $\lambda \neq \lambda_k$ without any assumptions on $q(x)$ and $r(x)$ except continuity as has been shown by analytic continuation in §7.

Combining (2) with (3), we find

$$\sum_{k=1}^{\infty}\left[\frac{\psi_k(x)\psi_k(y)}{(\lambda-\lambda_k)^2}\frac{\partial\lambda_k}{\partial a}+\frac{\psi_k(x)\dfrac{\partial\psi_k(y)}{\partial a}+\psi_k(y)\dfrac{\partial\psi_k(x)}{\partial a}}{\lambda-\lambda_k}\right]$$

$$= p(a)\left(\sum_{k=1}^{\infty}\frac{\psi_k'(a)\psi_k(y)}{\lambda-\lambda_k}\right)\left(\sum_{k=1}^{\infty}\frac{\psi_k(x)\psi_k'(a)}{\lambda-\lambda_k}\right)$$

Letting $\lambda \to \lambda_k$ and equating coefficients of $(\lambda-\lambda_k)^{-2}$ we find

$$\frac{\partial\lambda_k}{\partial a} = p(a)(\psi_k(a))^2. \tag{4}$$

By equating coefficients of $(\lambda-\lambda_k)^{-1}$ we find

$$\frac{\partial\psi_k(x)}{\partial a} = p(a)\psi_k'(a)\sum_{j\neq k}^{\infty}\frac{\psi_i(x)\psi_i'(a)}{\lambda_{1_k}-\lambda_i}. \tag{5}$$

We can also derive expressions for the variation of the sums

$$S_n(x, y, a) = \sum_{k=1}^{\infty}\frac{\psi_k(x)\psi_k(y)}{\lambda_k^n}, \quad (n=1, 2, 3, \ldots).$$

Since

$$\frac{-1}{\lambda-\lambda_k} = -\frac{1}{\lambda_k}-\frac{\lambda}{\lambda_k^2}-\frac{\lambda^2}{\lambda_k^3}-\cdots.$$

we have

$$R(x, y, \lambda, a) = -\sum_{n=1}^{\infty}S_n(x, y, a)\lambda^{n-1}.$$

Consequently,

$$\frac{\partial S_n}{\partial a}(x, y, a) = -p(a) \sum_{\substack{i+k=n+1 \\ 1 \le i \le n \\ 1 \le k \le n}} \frac{\partial S_i(a, y, a)}{\partial x} \cdot \frac{\partial S_k(x, a, a)}{\partial y} .$$

If the variational equations are to be used numerically some form of truncation must be used; see Chapter 14.

11. MATRIX CASE

Analogous results can be obtained for a Green's function associated with the vector-matrix system

$$\frac{d}{dt}\left(P(t)\frac{d}{dt}x(t)\right)+Q(t)x(t) = y(t),$$

$$a < t < 1, \quad x(a) = c, \quad x(1) + Bx(1) = 0, \qquad (1)$$

where P, Q, and B are $n \times n$ matrices and x, y, and c are n-dimensional vectors. We shall assume that P, Q, and y are continuous on the closed interval [a, 1] and that P(t) has an inverse for each t. Also we shall assume that P, Q, and P(I)B are symmetric matrices so that the system is self-adjoint.

For notation, we denote the transpose of a matrix A by A^* and we use (u, v) to denote the inner product of two vectors u and v.

Associated with this boundary problem there is the variational problem of maximizing J(x, y) where

$$J(x, y) = \int_a^1 \{(Qx, x)-(P\dot{x}, \dot{x})-2(y, x)\}dt-(P(1)Bx(1), x(1))$$

subject to the condition that x(a) = c. Let

$$f(a, c) = \max_{x(a)=c} J(x, y).$$

As in the scalar case, a vector x(t) which yields the maximum must be a solution of the boundary value problem (1).

Proceeding as in §3 one can use the principle of optimality from dynamic programming to derive the partial differential equation

$$-\frac{\partial f}{\partial a}(a,\ c) = \frac{1}{4}\left(p^{-1}(a)\frac{\partial f}{\partial c}, \frac{\partial f}{\partial c}\right) -2(c,\ y(a))+Q(a)c,\ c) \qquad (2)$$

where if $c_1, c_2, \ldots, c_n$ are the components of c then

$$\frac{\partial f}{\partial c} = \begin{bmatrix} \frac{\partial f}{\partial c_1} \\ \frac{\partial f}{\partial c_2} \\ \vdots \\ \frac{\partial f}{\partial c_n} \end{bmatrix}$$

We wish to use this partial differential equation to study the Green's function for the system (1). The Green's function is an $n \times n$ matrix $K(t, s)$ for which

$$x_0(t) = \int_a^1 K(t,\ s)y(s)\ ds \qquad (3)$$

is a solution of (1) with $x_0(a) = 0$. As before a solution of (1) with $x(a) = c$ can be written as a sum

$$x(t) = x_0(t) + \Phi(t)c$$

where $\Phi(t)$ is the $n \times n$ matrix function which is the solution of

$$\frac{d}{dt}(p(t)\frac{d}{dt}\Phi(t))+Q(t)\Phi(t)=0, \quad \begin{array}{l}\Phi(a)=1, \\ \Phi(1)+B\Phi(1)=0\end{array} \qquad (4)$$

Since $x(t)$ is the maximizing vector for the variational problem, we obtain

$$f(a, c) = -\int_a^1 (y(t), x(t))\,dt - (P(1)B(1)x(1), x(1)) +$$

$$+(P(a)\dot{x}(a), x(a)) - (P(1)\dot{x}(1), x(1)).$$

$$= -\int_a^1 (y(t), x_0(t))\,dt - \int_a^1 (y(t), \Phi(t)c)dt +$$

$$+(P(a)\dot{x}_0(a), c) + (P(a)\dot{\Phi}(a)c, c). \tag{5}$$

Also,

$$\int_a^1 (y(t), \Phi(t)c)dt = \int_a^1 (x_0(t), \frac{d}{dt} P(t)\dot{\Phi}(t)c + Q\Phi c)\,dt +$$

$$+ (P(1)\dot{x}_0(1), \Phi(1)c) - (P(a)\dot{x}_0(a), \Phi(a)c) - (x_0(1), P(1)\dot{\Phi}(1)c +$$

$$+ (x_0(a), P(a)\dot{\Phi}(a)c) = -(P(a)\dot{x}_0(a), c), (P(1)Bx_0(1), \Phi(1)c +$$

$$+ (P(1)x_0(1), B\Phi(1)c) = -(P(a)\dot{x}_0(a), c) \tag{6}$$

because of the assumption that $P(1)B$ is symmetric.

Hence by (3) and (5),

$$f(a, c) = -\int_a^1\int_a^1 (K(t, s)y(s), y(t))ds\,dt -$$

$$-2\int_a^1 (y(t), \Phi(t)c)dt + (P(a)\dot{\Phi}(a)c, c).$$

Also it is possible to express Φ in terms of a partial derivative of K, By (6).

$$\int_a^1 (c, \Phi^*(s)y(s))dt = -(c, P(a)\int_a^1 \frac{\partial K}{\partial t} (t, s, a)_{t=a} y(s)ds)$$

for all continuous vector functions y and all vectors c.
Hence,

$$\Phi^*(s) = -P(a) \frac{\partial K}{\partial t} (t, s, a)_{t=a}. \tag{7}$$

Next, we combine the above expression for f(a, c) with the partial differential equation (2) and equate the terms independent of c to obtain

$$\int_a^1\int_a^1 \left(\frac{\partial K(t, s, a)}{\partial a} y(s), y(t)\right)ds\ dt =$$

$$= (P^{-1}(a)\int_a^1 \Phi^*(t)y(t)\,dt, \quad \int_a^1 \Phi^*(t)y(t)\,dt),$$

or

$$\int_a^1\int_a^1 \left(\left\{\frac{\partial K}{\partial a}(t, s, a) - \Phi(t)P^{-1}(a)\Phi^*(s)\right\}y(s), y(t)\right)ds\ dt=0. \tag{8}$$

Now we "equate coefficients" to conclude that

$$\frac{\partial K}{\partial a}(t, s, a) = \Phi(t)P^{-1}(a)\Phi^*(s). \tag{9}$$

This formal argument can be justified by the following result corresponding to the lemma proved in §5.

LEMMA. Let M(t, s) be a continuous n × n matrix function for $a \le s \le 1$, $a \le t \le 1$ and suppose that M has the symmetry property

$$M^*(t, s) = M(s, t).$$

If

$$\int_a^1\int_a^1 (M(t, s)y(s), y(t))ds\ dt = 0$$

for all continuous vector functions y, then $M(s, t) \equiv 0$.

We omit a proof of this lemma because one can be constructed in a way quite similar to that employed in §5. It is also not difficult to verify that the matrix function in (8) has the required symmetry property because the system is self-adjoint.

Combining (7) and (9) we obtain

$$\frac{\partial K}{\partial a}(t, s, a) = \frac{\partial K}{\partial s}(t, s, a)_{t=a} P(a) \frac{\partial K}{\partial t}(t, s, a)_{t=a}. \quad (10)$$

The problem we have just treated does not contain as a special case the problem with the boundary condition $x(1) = 0$. However, only a small change is required to handle this problem by the same method. In the definition of $J(x, y)$ one can omit the term $-(P(1)B(1)x(1), x(1))$ and for the maximization problem add the constraint $x(1) = 0$. The remainder of the argument is quite similar and the final result obtained, equation (10), is the same.

The approach by way of change of variable as in §6 can also be followed in the matrix case. Set

$$x(t) = (I+(1-t)B)(I+(1-a)B)^{-1}c+w(t).$$

This transformation has the property that if $x(a)=c$ then $w(a)=0$; and if $\dot{x}(1)+Bx(1)=0$, then $w(1)+Bw(1)=0$. Here we must also make the additional assumption that the inverse $(I+(1-a)B)^{-1}$ exists. Proceeding as in §6, we obtain the equation

$$\Phi(t) = (I+(1-t)B-\int_a^1 K(t, s)[Q(s)(I+(1-s)B) -$$

$$- \frac{d}{ds}(P(s))B]ds)(I+(1-a)B)^{-1},$$

which allows the derivation of another expression for $\partial K/\partial a$.

12. INTEGRAL EQUATIONS

Let us now indicate briefly how the same formalism may be applied to integral equations. The equation

$$u(x) = v(x) + \lambda\int_a^1 k(x, y)u(y)dy \quad (1)$$

has the solution

$$u(x) = v(x) + \int_a^1 K(x, y, \lambda)v(y)dy, \quad (2)$$

where $K(x, y, \lambda)$ is the Fredholm resolvent.

If $k(x, y)$ is positive definite, we can consider (1) to be the Euler-Lagrange equation corresponding to the problem of minimizing the quadratic functional

$$Q(u) = \lambda \int_a^1 \int_a^1 k(x, y)u(x)u(y)dx\, dy + 2\int_a^1 u(x)v(x)dx - \int_a^1 u^2(x)dx. \tag{3}$$

Regarding the minimum over u as a function of a and a functional of v, we can proceed very much as above. Continuing as in §3.4, we can derive variational equations for the characteristic values and functions.

BIBLIOGRAPHY AND COMMENTS

Section 1. We are following the paper:

R. Bellman and S. Lehman, 'Functional Equations in the Theory of Dynamic Programming - X: Resolvents, Characteristic Functions and Values', Duke Mathematical Journal 27 (1960), 55-70.

Section 7. See also:

R. Bellman and S. Lehman, 'Functional Equations in the Theory of Dynamic Programming - IX: 'Variational Analysis, Analytic Continuation, and Imbedding of Operators', Proceedings of the National Academy of Sciences 44 (1958), 905-907.

Section 12. See

R. Bellman, 'Functional Equations in the Theory of Dynamic Programming - VII: A Partial Differential Equations for the Fredholm Resolvent', Proceedings of the American Mathematical Society 8 (1957) 435-440.

The equation for the resolvent kernel, an equation of Riccati type, has been used extensively by Ueno in his work on radiative transfer.

Chapter VII

THE HADAMARD VARIATIONAL FORMULA

1. INTRODUCTION

In an earlier chapter, the functional equation technique of dynamic programming was applied to obtain a variational equation for a Green's function corresponding to a second order ordinary differential equation. In the present chapter, this method is extended to apply to elliptic partial differential operators, and a first consequence is the classical Hadamard variational formula.

The technique presented here utilizes the principle of optimality in the following fashion. Given a one-parameter family of regions, monotone under inclusion, one takes the minumum value of a certain integral on any given region R, subject to certain restrictions, to be a functional of those restrictions and the region R. Then if $R^* \subset R$ the functional on R can be approximated by means of a related functional on R^* satisfying slightly different restrictions. This leads to a Gateaux difference equation from which one easily derives the Hadamard relation.

The method is initially presented for the Laplace operator on R, and appropriate generalizations are indicated in §6 and §7.

The same methods may be used to handle multidimensional variational problems.

2. PRELIMINARIES

Let R be a bounded connected region of n-dimensional real Euclidean space, whose boundary ∂R is of class C_2. For convenience we shall not explicitly write out the differentials of volume and surface area in integrals over R and ∂R. Given any twice-differentiable function u on R, let Δu and u_p represent the Laplacian of u and the restriction of u to its limiting values on ∂R, respectively. Then, if v and w are suitable functions on R and ∂R, the boundary value problem

$$\Delta u = v, \qquad u_p = w \tag{1}$$

possesses the unique solution

$$u(p) = \int_{q \in R} g(p, q)v(q) + \int_{q \in \partial R} g_n(p, q)w(q) \tag{2}$$

where g is the Green's function for R normalized by the condition that

$$\int_{q \in S} \Delta g(p, q) = \int_{q \in \partial S} g_n(p, q) = 1. \tag{3}$$

Here S is any sphere with center p which lies inside R, and g_n is the exterior normal derivative of G on ∂S. The two integrals in (2) represent the solutions $u^{(1)}$ and $u^{(2)}$ of the boundary value problems

$$\Delta u = v, \qquad u_p = 0 \tag{4}$$

and

$$\Delta u = 0, \qquad u_p = w, \tag{5}$$

respectively, and they are orthogonal in the sense that

$$\int_R \nabla u^{(1)} \cdot \nabla u^{(2)} = \int_{\partial R} u_p^{(1)} u_n^{(2)} - \int_R u^{(1)} \Delta u^{(2)} = 0, \tag{6}$$

where $\nabla u^{(i)}$ is the gradient of $u^{(i)}$, i=1,2.

3. A MINIMUM PROBLEM

Among those functions u such that $u_p = 0$, the function $u^{(1)}$ maximizes the integral

$$\int_{p \in R} \int_{q \in R} g(p, q)(\Delta u - v)(p)(\Delta u - v)(q).$$

Hence, since the maximum value is zero, and since

$$\int_{q \in R} g(p, q)\Delta u(q) = u(p) \tag{1}$$

for any function u such that $u_p = 0$, one obtains an extremal condition

$$\min_{u|u_p=0} \int_R 2uv+|\nabla u|^2 = \min_{u|u_p=0} \int_R (2v-\Delta u)u$$
$$= \int_{p\in R}\int_{q\in R} g(p, q)v(p)v(q) \tag{2}$$

with the minimizing function $u^{(1)}$. Define

$$f(v, w) = \min_{u|u_p=w} \int_R [2uv + |\nabla u|^2] \tag{3}$$

so that

$$f(v, 0) = \int_{p\in R}\int_{q\in R} g(p, q)v(p)v(q). \tag{4}$$

It should be noted that the first equality in (2) fails for those u such that $u_p = w \neq 0$.

Suppose that $u^{(2)}$ is given as the solution of (2.5). Then (3) may be rewritten, by means of (2.6) as

$$f(v, w) = \min_{u|u_p=0} \int_R [2(u+u^{(2)})v + |\nabla u+\nabla u^{(2)}|^2]$$
$$= \min_{u|u_p=0} \int_R [2uv+|\nabla u|^2] + \int_R [2u^{(2)}v+|\nabla u^{(2)}|^2]$$
$$= f(v, 0)+ 2\int_R u^{(2)}v + \int_R |\nabla u^{(2)}|^2. \tag{5}$$

In particular, writing εw in place of w,

$$f(v, \varepsilon w) = f(v, 0) + 2\varepsilon \int_R u^{(2} v + \varepsilon^2\int_R |\nabla u^{(2)}|^2. \tag{6}$$

Since $u^{(2)}$ is known explicitly in terms of w, this enables us to compute the Gateaux difference

$$f(v, \varepsilon w)-f(v, 0)=2\varepsilon \int\int_{\substack{p\in R\\ q\in\partial R}} g_n(p, q)v(p)w(q)+o(\varepsilon). \tag{7}$$

4. A FUNCTIONAL EQUATION

Let φ be a non-negative function of class C_2 on ∂R, and let ∂R^* be the surface obtained from ∂R by a displacement δn along the interior normal, where $\delta n = \varepsilon\varphi$. If u is any differentiable function on R such that $u_p = 0$, then the restriction u_{p^*} of u to ∂R^* is $-u_n^p \delta n + o(\varepsilon)$. We extend the definition of f to the class of regions R^* with boundaries ∂R^* by setting

$$f(\varepsilon, v, w) = \min_{u|u_{p^*}=w} \int_{R^*} [2uv + |\nabla u|^2]. \tag{1}$$

If $u_p = 0$ then $|\nabla u|^2 = u_n^2$ on ∂R, so that the n-dimensional analog of the principle of optimality implies

$$f(0, v, 0) = \min_{u_n}[f(\varepsilon, v, - u_n \delta n) + \int_{\partial R} \delta n u_n^2 + o(\varepsilon)]. \tag{2}$$

Set $\delta f(v, w) = f(\varepsilon, v, w) - f(0, v, w)$ and note that

$$\delta f(v, - u_n \delta n) = \delta f(v, 0) + o(\varepsilon). \tag{3}$$

In this notation one may apply (3.7) and (2) to obtain

$$\min_{u_n}\left[\delta f(v, 0) - 2\int_{p \in R} \int_{q \in \partial R} g_n(p, q) v(p) \delta n(q) u_n(q) + \int_{\partial R} \delta n u_n^2\right] = o(\varepsilon). \tag{4}$$

The Euler-Lagrange equation of (4) is

$$u_n(q) = \int_{p \in R} g_n(p, q) v(p), \tag{5}$$

and it follows that

$$\delta f(v, 0) = \int_{s \in \partial R} \int_{p \in R} \int_{q \in R} \delta n(s) g_n(p, s) g_n(q, s) v(p) v(q) + o(\varepsilon) \tag{6}$$

5. THE HADAMARD VARIATION

Let $g(\varepsilon, p, q)$ represent the Green's function of the region R^*, and let $\delta g(p, q) = g(\varepsilon, p, q) - g(0, p, q)$. We wish to derive the Hadamard relation between δg and δn. For the region R^* (3.4) becomes

$$f(\varepsilon, v, 0) = \int_{p \in R^*} \int_{q \in R^*} g(\varepsilon, p, q)v(p)v(q) \qquad (1)$$

and since g vanishes and possesses a bounded normal derivative on ∂R it follows that

$$\delta f(v, 0) = \int_{p \in R} \int_{q \in R} \delta g(p, q)v(p)v(q)+o(\varepsilon). \qquad (2)$$

Since v is arbitrary, (4.6) and (2) together imply

$$\delta g(p, q) = \int_{s \in \partial R} \delta n(s)g_n(p, s)g_n(q, s)+o(\varepsilon). \qquad (3)$$

which is Hadamards relation.

The preceding derivation is valid only when $R^* \subset R$. To prove (3) in general it suffices to consider R and R^* as regions both interior to a third region $\bar{R}$, and to consider the difference of the variation of $\bar{R}$ to R^* and the variation $\bar{R}$ to R. This device is due to Hadamard and is also applied in the standard derivative of (3).

6. LAPLACE-BELTRAMI OPERATOR

The Hadamard relation remains valid if Δ is replaced by any other self-adjoint second order differential operator which possesses a Green's function g. Thus, for example, Δ may be replaced by an arbitrary Laplace-Beltrami operator merely by furnishing R with an appropriate Reimannian metric. In this case, there is no change in the preceding derivation.

7. INHOMOGENEOUS OPERATOR

Alternatively we may add a multiplication to obtain the operation $u \to \Delta u + \alpha(p)u$. Assuming that $\alpha(p)$ is sufficiently small, we may again consider the functional f defined by

$$f(v, w) = \min_{u|u_p=w} \int_R [2uv + |\nabla u|^2 - \alpha u^2]. \qquad (1)$$

The appropriate orthogonality relation is now

$$\int_R [\nabla u^{(1)} \cdot \nabla u^{(2)} - \alpha u^{(1)} u^{(2)}] = \int_{\partial R} u_p^{(1)} u_n^{(2)}$$

$$- \int_R u^{(1)} [\Delta u^{(2)} + \alpha u^{(2)}] = 0 \tag{2}$$

The remainder of the argument proceeds as before.

Since the variational formula is independent of α, one may conclude that it is valid wherever the Green's function exists.

BIBLIOGRAPHY AND COMMENTS

Section 1. These results were given in:

R. Bellman and H. Osborn, 'Dynamic Programming and the Variation of Green's Functions'. Journal of Mathematics and Mechanics 7 (1958), 81-85.

The one-dimensional variational problem was treated in:

S.E. Dreyfus, Dynamic Programming and the Calculus of Variations, Academic Press, Inc., New York, 1965.

Chapter VIII

THE TWO-DIMENSIONAL POTENTIAL EQUATION

1. INTRODUCTION

In this chapter we wish to discuss some aspects of the potential equation

$$u_{xx} + u_{yy} = 0, \qquad (x, y) \in R, \tag{1}$$

$$u=g(x, y), \quad (x, y) \in \Gamma, \tag{2}$$

where Γ is the boundary of the region R in figure 1. We want to consider its connection with the minimization of the quadratic functional

$$D(u) = \int_R (u_x^2 + u_y^2)dR, \tag{3}$$

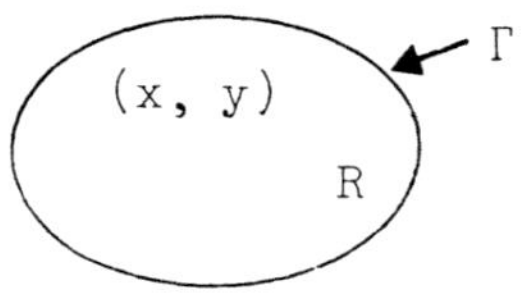

Figure 1

The minimum of the quadratic functional is itself a quadratic functional. If we discretize, this minimum is a quadratic form. It is this fact which enables us to overcome the dimensionality problem.

2. THE EULER-LAGRANGE EQUATION

We can readily show that the potential equation is the variational equation associated with D(u). Let v=0 on Γ so that the functions u and u+v satisfy the same boundary conditions on Γ. Then

$$D(u+v) = D(u)+D(v)+2\int_R (u_x v_x + u_y v_y)dR. \qquad (1)$$

Applying Green's theorem, we see that the third term vanishes if u satisfies the potential equation. Hence if u satisfies (1.1) we have

$$D(u+v) > D(u) \qquad (2)$$

for any nontrivial v. Thus, if (1.1) and (1.2) possess a solution, they possess a unique solution, since any solution minimizes D(u).

Conversely, it can be shown from first principles that D(u) possesses a minimum over the class of functions u such that $u_x, u_y \in L^2$ with u satisfying the specified boundary condition. Further it can be shown that the minimizing function is determined by (1.1) and (1.2).

Thus the problem of solving the potential equation and minimizing the Dirichlet functional D(u) are equivalent. We shall focus henceforth on the minimization problem.

3. INHOMOGENEOUS AND NONLINEAR CASES

It follows from the foregoing that the problem of solving the inhomogeneous problem

$$u_{xx} + uu_{yy} = h(x, y), \quad u(x, y) = 0, \quad (x, y) \in \Gamma \qquad (1)$$

and minimizing

$$D_1(y) = \int_R (u_x^2 + u_y^2 + 2h(x, y)u)dR, \qquad (2)$$

subject to u=0 on Γ, $u_x, u_y \in L^2(R)$, are equivalent.

Similarly, the problem of solving

$$u_{xx} + u_{yy} + k(x, y)u = 0, \qquad (3)$$

and minimizing

$$D_2(u) = \int_R (u_x^2 + u_y^2 - k(x, y)u^2)dR, \qquad (4)$$

are equivalent provided that $D_2(u)$ is a positive-definite functional. A sufficient condition for this $\max_R |k(x, y)| < \lambda_1$, where λ_1 is the smallest characteristic value of the problem

$$u_{xx} + u_{yy} + \lambda_1 u = 0, \quad u=0, \quad (x, y) \in \Gamma.$$

Finally, the nonlinear equation

$$u_{xx} + u_{yy} - h(u) = 0 \tag{5}$$

may be associated with the functional

$$\int_R (u_x^2 + u_y^2 + 2g(u))dR. \tag{6}$$

where $g(u) = h(u)$.

4. GREEN'S FUNCTION

The solution of the inhomogeneous problem

$$u_{xx} + uu_{yy} = h(x, y), \quad u=0, \quad (x, y) \in \Gamma, \tag{1}$$

may be expressed in the form

$$u = \int_R k(x, y, x_1, y_1)h(x_1, y_1)dR. \tag{2}$$

The kernel k is called the Green's function associated with the particular equation and boundary condition.

For the foregoing case we have $k \leq 0$, a result we wish to use subsequently. Let us derive this fundamental property directly from the associated variational problem.

5. TWO-DIMENSIONAL CASE

In chapter 6, we consider the one-dimensional case. The same method of proof can be used in higher dimensional cases. Let

$$D(u) = \int_R [u_x^2 + u_y^2 - q(x, y)u^2 + 2h(x, y)u]dR \tag{1}$$

be the quadratic function associated with the equation

$$u_{xx} + u_{yy} + q(x, y)u = h(x, y), \quad u=0, \quad (x, y) \in \Gamma \tag{2}$$

We suppose that the quadratic part of D(u) is positive definite.

Under this assumption we wish to show that $h \geq 0$ implies that $u \leq 0$. We proceed by contradiction. Let $u \geq 0$ in $R_1 - R$ and consider the new function

$$\begin{aligned} v &= -u \quad p \in R_1, \\ v &= u, \quad p \in R-R_1. \end{aligned} \tag{3}$$

As before, we see that $D(v) < D(u)$, contradiction. Thus $h \geq 0$ implies $u \leq 0$, whence as before the Green's function is nonpositive.

6. DISCRETIZATION

One powerful approach to the study of the properties of differential equations is the use of associated difference equations. This procedure is particularly useful numerically when we contemplate the use of analog or digital computers. However, it is equally valuable for theoretical purposes. The fundamental observation is that the approximation

$$\begin{aligned} u_{xx}(x, y) &\cong \frac{u(x+\Delta, y)+u(x-\Delta, y)-2u(x, y)}{\Delta^2}, \\ u_y(x, y) &\cong \frac{u(x, y+\Delta)+u(x, y-\Delta)-2u(x, y)}{\Delta^2}, \end{aligned} \tag{1}$$

converts the potential equation into a system of linear algebraic equations. This transformation, of course, is the beginning of the real problem, that of obtaining useful analytic and computational results from the linear system.

7. RECTANGULAR REGION

Let us see what this involves for a rectangular region (figure 2). Let integers M and N be chosen and the positive quantities Δ and δ be determined by $M\Delta = a$, $N\delta = b$. Let

$$u_{mn} = u(m\Delta, n\delta), \quad m=0,1, \ldots, M, \quad n=0,1, \ldots, N. \tag{1}$$

The original boundary condition,

$$u(x, y) = g(x, y), \quad (x, y) \in \Gamma, \tag{2}$$

yields the corresponding boundary conditions

$$
\begin{aligned}
u_{0n} &= g(0,\ n\delta), \quad n=0,\ 1,\ \ldots,\ N \\
u_{Mn} &= g(a,\ n\delta), \\
u_{m0} &= g(m\Delta,\ 0), \quad m=0,\ 1,\ \ldots,\ M, \\
u_{mN} &= g(m\Delta,\ b)
\end{aligned}
\tag{3}
$$

We suppose that $g(x, y)$ is continuous so that $g(0, 0)$ and $g(m\Delta, N\delta)$ are unambiguously defined by the foregoing

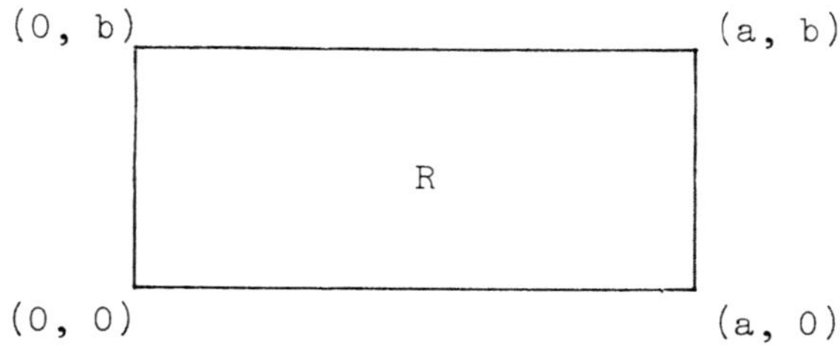

Figure 2

Upon setting $x = m\Delta$, $y = n\delta$, the potential equation yields the set of relations,

$$
\frac{u((m+1)\Delta,\ n\delta) + u((m-1)\Delta,\ n\delta) - 2u(m\Delta,\ n\delta)}{\Delta^2} + \frac{u(m\Delta,(n+1)\delta) + u(m\Delta,(n-1)\delta) - 2u(m\Delta,\ n\delta)}{\delta^2} \cong 0. \tag{4}
$$

Hence we obtain the linear difference equations

$$
\frac{v_{m+1,\ n} + v_{m-1,\ n} - 2v_{mn}}{\Delta^2} + \frac{v_{m,\ n+1} + v_{m,\ n-1} - 2v_{mn}}{\delta^2} = 0,
$$
$$
m = 0,\ 1,\ \ldots,\ M, \qquad n = 0,\ 1,\ \ldots,\ N. \tag{5}
$$

These relations, together with (3), constitute a system of linear algebraic equations for the quantities v_{mn}.

8. ASSOCIATED MINIMIZATION PROBLEM

It is necessary to consider the existence and uniqueness of the solution of the linear system in (7.5). We can answer these questions readily by noting that the linear difference equations are the variational equations associated with the quadratic form

$$Q_{M,N}(v) = \sum_{m,n}\left[\left(\frac{v_{M+1,n}-v_{m,n}}{\Delta}\right)^2+\left(\frac{v_{m,n+1}-v_{m,n}}{\delta}\right)^2\right], \qquad (1)$$

where m ranges over 0, 1, 2, ..., M and n over 0, 1, 2, ..., N. The boundary values are determined as in (7.3). Since $Q_{M,N}(v)$ is clearly of positive-definite nature, it possesses a unique minimum value, attained by the unique solution of (7.5).

9. APPROXIMATION FROM ABOVE

The foregoing minimization problem is obtained from the original minimization problem, (associated with D(u)), by restricting attention to functions u for which u_x are constant over the rectangles

$$m\Delta \le x \le (m+1)\Delta, \qquad n\delta \le y \le (n+1)\delta. \qquad (1)$$

It follows that the minimum over this subclass of functions is greater than or equal to the minimum over the original class of functions such that u_x and u_y are in $L^2(R)$. Hence, for any $\Delta, \delta > 0$, we have

$$\min_v Q_{M,N}(v) \ge \min_u D(u). \qquad (2)$$

10. DISCUSSION

If we wish to obtain an accurate estimate for $u(m\Delta, n\delta)$ in the foregoing fashion, the quantities Δ and δ must be taken small. If we do this, however, we are left with a high order system of linear algebraic equations. The task of solving this system numerically is an onerous and unreliable one if straightforward techniques are employed.

Fortunately, the special structure of the system allows the use of some powerful and ingenious techniques in a number of cases. These methods, as all methods, possess advantages and disadvantages. We shall pursue a quite different approach here.

11. SEMIDISCRETIZATION

It is occasionally advantageous to introduce discretization in one variable only. Thus, for example, we may wish to use a grid only in the y-direction. The potential equation is then replaced by

$$W_{xx}^{(m)} = \frac{W^{(m-1)} + W^{(m+1)} - 2W^{(m)}}{\Delta^2}, \tag{1}$$

where $W^{(m)} = W(x, m\Delta)$.

For the region R_1, where $0 \le x \le a$, $0 \le y \le b$, the integer m ranges from 1 to M-1 with the two-point boundary conditions

$$W^{(m)}(0) = a_m, \quad W^{(m)}(a) = b_m, \quad W^{(0)}(x) = g(x), \quad W^{(M)}(x) = h(x),$$
$$m=0, 1, \ldots, M, \quad 0 \le x \le a. \tag{2}$$

The quantities a_m, b_m, $g(x)$ and $h(x)$ are immediately determined from (7.2).

The solution of linear differential equations of this nature can be effected by means of the solution of M-dimensional linear algebraic systems. The associated variational problem is the minimization of

$$\int_0^a \left[\sum_{m=1}^{M-1} (W_x^{(m)})^2 + \sum_{m=1}^{M-1} \left(\frac{W^{(m+1)} - W^{(m-1)}}{\Delta}\right)^2\right] dx. \tag{3}$$

The pcsitive-definite nature of this functional ensures the existence and uniqueness of the solution of (1) and (2).

12. SOLUTION OF THE DIFFERENCE EQUATIONS

The linear difference (or finite difference) equations, (7.5), can be written in matrix-vector form

$$Av = b, \tag{1}$$

where v is the (M-1)(N-1)-dimensional vector of the unknown $\{v_{ij}\}$ at the interior points

$$v = \begin{bmatrix} v_{1,\,1} \\ \vdots \\ v_{1,\,N-1} \\ v_{2,\,1} \\ \vdots \\ v_{M-1,\,N-1} \end{bmatrix} \tag{2}$$

The matrix A is determined by the finite difference approximation (7.4) and the grid spacing; the vector b is determined by the boundary conditions. The structure of A has many special properties. At the moment we again note that A is nonsingular which allows us to write v as

$$v = A^{-1}b. \tag{3}$$

Comparing this result with (4.2) it is clear that A^{-1} is the discrete analog of the Green's function. All the elements of A^{-1} are positive and thus the discrete Green's function possesses a property analogous to the negativity of $k(t, t_1)$. The proof follows the lines of the proof for the continuous case.

Since the dimension of A is large when an accurate numerical result is desired, a solution of (1) by conventional direct methods, such as Gaussian elimination, is impractical. Subsequently we will employ the block tri-diabonal structuré of A to our advantage. Going back to (7.4), we see that a row of A can have at most five nonzero entries. Due to sparseness, iterative methods can be effectively employed to solve (1) numerically.

13. THE POTENTIAL EQUATION

Let us now present an approach to the numerical solution of some types of elliptic equations based on the theory of dynamic programming. We start with the problem introduced above, the potential equation

$$\nabla^2 u = u_{xx} + u_{yy} = 0, \tag{1}$$

over a region R with the boundary conditions

$$u(x, y) = f(x, y), \tag{2}$$

on Γ the boundary of R.

We have seen that (1) is the Euler-Lagrange equation associated with the minimization of the quadratic functional

$$D(u) = \int\int_R [u_x^2 + u_y^2]\, dR, \tag{3}$$

subject to (2).

For initial purposes of exposition we will assume that R is the rectangle $0 \le x \le a$, $0 \le y \le b$.

Subsequently, we will treat some irregular regions as combinations of rectangular regions.

14. DISCRETIZATION

Our method will be to determine the minimum of a discretized version of the Dirichlet functional corresponding to the original potential equation. We will assume for convenience that a and b are such that

$$a = Nh, \quad b = Mh, \tag{1}$$

where M and N are integers. The modifications necessary if this is not the case are simple and do not effect the subsequent development. We seek a function at the points of a finite grid (or mesh). Specifically, we are interested in the points

$$u_{ij} = u(ih, jh), \quad i=0, 1, \ldots, N, \quad j=0, 1, \ldots, M. \tag{2}$$

Since we want a solution defined only at these $(N+1)(M+1)$ points, we would like to replace the partial derivatives at these points by terms involving only the function values at these points. Using a Taylor series expansion we have

$$\frac{\partial u_{ij}}{\partial x} = \frac{u_{ij}-u_{i-1,\, j}}{h} + O(h), \quad \frac{\partial u_{ij}}{\partial y} = \frac{u_{ij}-u_{i,\, j-1}}{h} + O(h), \tag{3}$$

a standard finite difference approximation.

Thus, to obtain a discrete version of the functional in (13.3) we replace the partial derivatives appearing by the expressions in (3) and assume that the derivatives are constant between grid points. Equivalently, we could apply rectangular integration to (13.3) and then use (3). Either way the double integral in the Dirichlet functional can be approximated by the double sum

$$J(u) = \sum_{i=1}^{N} \sum_{j=1}^{M} [(u_{ij} - u_{i,j-1})^2 + (u_{ij} - u_{i-1,j})^2] \tag{4}$$

We can easily verify that the discretization error is $O(h^2)$. The discretized functional is subject to the original boundary conditions and thus the values u_{0j}, u_{i0}, u_{Nj}, and u_{iM} are determined by (13.2).

We are left with the finite dimensional problem of finding

$$\min_{(u_{ij})} J(u), \qquad i=1,2,\ldots,N-1, \quad j=1,2,\ldots,M-1. \tag{5}$$

It should be clear that since we are interested only in the minimizing $\{u_{ij}\}$ and not the minimum value of J(u) itself we can remove constant terms from (4) without changing the set of minimizing $\{u_{ij}\}$. It will turn out to be convenient to remove from (4) all terms of the form $(u_{iM} - u_{i-1,M})^2$ which are fixed by the boundary conditions. In this manner we replace (4) by the more convenient functional

$$J(u) = \sum_{i=1}^{N} \left[\sum_{j=1}^{M} (u_{ij}-u_{i,j-1})^2 + \sum_{j=1}^{M-1} (u_{ij}-u_{i-1,j})^2 \right]. \tag{6}$$

15. MATRIX-VECTOR FORMULATION

Although we have now completely discretized the original problem, it is still not in a form which can easily be solved. We will first put the problem into matrix-vector form and then convert it into a multistage decision process which can be handled by dynamic programming.

Since the functional is a quadratic form, we can conveniently rewrite the problem using inner product notation. First, we define the vectors of interior values

$$u_R = \begin{bmatrix} u_{R1} \\ u_{R2} \\ \vdots \\ u_{R,\ M-1} \end{bmatrix}, \quad R=1,\ 2,\ \ldots,\ N-1, \tag{1}$$

and note that the vectors u_0 and u_N are given by the boundary conditions. Now we define a symmetric matrix of order M-1, Q; a set of (M-1)-dimensional vectors, r_R, and scalars, s_R, as follows

$$Q = (q_{ij}) \quad \text{where} \quad q_{ij} = \begin{cases} 2, & i=j, \\ -1, & |i-j| = 1, \\ 0, & \text{otherwise}, \end{cases}$$

$$r_R = [r_{RJ}] \quad \text{where} \quad r_{Rj} = \begin{cases} u_{R0}, & j=1, \\ u_{RM}, & j=M-1 \\ 0, & \text{otherwise}, \end{cases} \tag{2}$$

$$s_R = u_{R0}^2 + u_{RM}^2.$$

Clearly, Q is a fixed matrix, while r_R and s_R are functions of only the boundary conditions. With this notation, J(u) becomes, in inner product form

$$J(u) = \sum_{R=1}^{N} [(Q_{uR}, u_R)-(2r_R, u_R)+s_R+(u_R-u_{R-1}, u_R-u_{R-1})]. \tag{3}$$

16. DYNAMIC PROGRAMMING

Instead of the original problem, let us consider the sequence of variational problems

$$f_R(v) = \min_{[u_R,\ u_{R+1},\ \ldots,\ u_{N-1}]} \sum_{i=R}^{N} [(Qu_i, u_i)-(2r_i, u_i)+s_i. + \\ + (u_i-u_{i-1}, u_i-u_{i-1})], \quad R=1,\ 2,\ \ldots,\ N-1, \tag{1}$$

where v is defined as

$$v = u_{R-1}. \tag{2}$$

Clearly, by comparison with (15.3) we have

$$f_1(u_0) = \min_{\{u_R\}} J(u). \tag{3}$$

Each of the problems in (1) can be regarded as the solution of the potential equation over a truncated rectangle with the left boundary condition v, unspecified as in Figure 3. This is our fundamental imbedding.

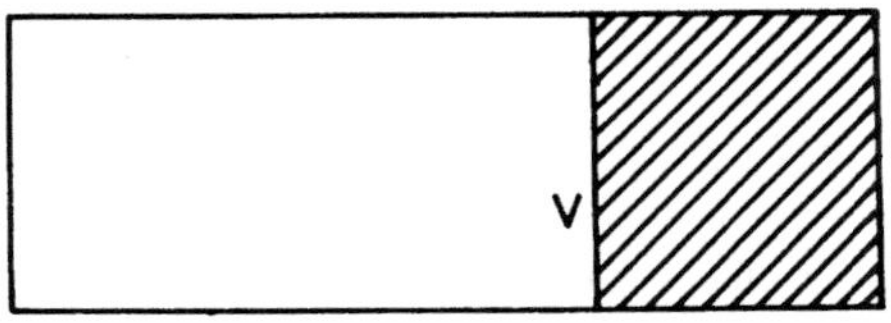

Figure 3

Let us next find a relationship between the function $f_R(v)$ and $f_{R+1}(v)$. Noting that only the first four terms of (1) depend on u_R, we can write

$$f_R(v) = \min_{[u_R,\, u_{R+1},\, \dots,\, u_{N-1}]} [(Qu_R, u_R) - 2(r_R, u_R) + s_R + (u_R - v, u_R - v) + \sum_{i=R+1}^{N} \{(Qu_i, u_i) - 2(r_i, u_i) + s_i + (u_i - u_{i-1}, u_i - u_{i-1})\}]. \tag{4}$$

The minimization over $u_{R+1}, u_{R+2}, \dots, u_{N-1}$ can be moved past the first four terms, resulting in the expression

$$f_R(v) = \min_{u_R} \Big\{(Qu_R, u_R) - 2(r_R, u_R)+s_R+(u_R-v, u_R-v)+$$

$$+ \min_{u_{R+1}, u_{R+2}, \ldots, u_{N-1}} \Big[\sum_{i=R+1}^{N} (Qu_i, u_i)-2(r_i, u_i)+$$

$$+ s_i + (u_i-u_{i-1}, u_i-u_{i-1})\Big]\Big\} \qquad (5)$$

However, by comparing the terms to be minimized over $u_{R+1}, u_{R+2}, \ldots, u_{N-1}$ with the definition in (1) of $f_R(v)$ we see that (5) can be rewritten as

$$f_R(v) = \min_{r_R}[(Qu_R, u_R)-2(r_R, u_R)+s_R+(u_R-v, u_R-v)+f_{R+1}(u_R)] . \qquad (6)$$

This result may also be considered to be a consequence of the principle of optimality. The choice of u_R must balance the cost of the present stage, the first four terms of (5), against the cost of continuing the process starting with u_R, and proceeding through N-R additional stages.

Since u_N is fixed by the boundary conditions, we have

$$f_N(v) = (Qu_N, u_N)-2(r_N, u_N)+s_N+(u_N-v, u_N-v). \qquad (7)$$

Theoretically, we could now solve for the minimizing $\{u_R\}$ by starting with (7) and solving (6) backwards until $f_1(u_0)$ is determined. However, we have yet to exploit fully the quadratic nature of the variational problem. When we do so, the analytic and computational features will simplify considerably.

17. RECURRENCE EQUATIONS

It can easily be verified inductively that $f_R(v)$ must be quadratic in v,

$$f_R(v) = (A_R v, v) - 2(b_R, v) + c_R, \qquad (1)$$

where A_R is a symmetric matrix. Furthermore the quantities A_R, b_R, and c_R are independent of v. Substituting this quadratic form for $f_{R+1}(u_R)$ in (16.5) we have

$$f_R(v) = \min_{u_R} [(Qu_R, u_R) - 2(r_R, u_R)+s_R+(u_R-v, u_R-v) +$$

$$+ (A_{R+1}u_R, u_R)-2(b_{R+1}, u_R)+c_R]. \qquad (2)$$

The expression can be readily differentiated in order to find the minimizing u_R. The fact that this u_R actually yields a minimum can be easily verified from the positive definiteness of the functional. The minimizing u_R is found to be

$$u_R = [Q+A_{R+1}+I]^{-1}(v+b_{R+1}+r_R). \qquad (3)$$

Upon using this expression for u_R in (2) we obtain

$$f_R(v) = ([I - [I+Q+A_{R+1}]^{-1}]v, v) -$$

$$- (2[I+Q+A_{R+1}]^{-1}((b_{R+1}+r_R), v)+c_{R+1}+s_{R+1} -$$

$$- ([I+Q+A_{R+1}]^{-1}((b_{R+1}+r_R), b_{R+1}+r_R)). \qquad (4)$$

Thus by comparison with (1), we obtain the relations

$$A_R = I-[I+Q+A_{R+1}]^{-1},$$

$$b_R = [I+Q+A_{R+1}]^{-1}(b_{R+1}+r_R),$$

$$= [I-A_R](b_{R+1}+r_R), \qquad (5)$$

$$c_R = c_{R+1}+s_R-([I+Q+A_{R+1}]^{-1}(b_{R+1}+r_R, b_{R+1}+r_R)$$

$$= c_{R+1}+s_R-(b_R, b_{R+1}+r_R).$$

From (16.7) we get the initial conditions

$$A_N = I, \quad b_N = u_N, \quad c_N = ([I+Q]u_N, u_N)-2(r_N, u_N)+s_N. \qquad (6)$$

Since v is by definition u_{R-1}, (3) becomes

$$u_R = [I+Q+A_{R+1}]^{-1}(u_{R-1}+b_{R+1}+r_R), \qquad (7)$$

or

$$u_R = [I-A_R]u_{R-1}+b_R. \qquad (8)$$

Since we are primarily interested in u_R, we can ignore the equation for c_R.

Let us note that this argument may be used for inhomogeneous regions.

18. THE CALCULATIONS

The calculations proceed as follows. Equations (17.5) are solved successively starting with (17.6) until A_1 and b_1 have been determined. All intermediate values of A_R and b_R are stored. Relation (17.8) is an initial value problem starting with u_0, the second boundary condition, A_1 and b_1. This equation is iterated using the last computed value u_R and the stored values of A_R and b_R. As is typical of the dynamic programming approach, a two-point boundary value problem has been converted to two initial value problems.

19. IRREGULAR REGIONS

It is clear that the foregoing methods can be used for circular and elliptical regions. They can also be applied to triangular and more generally trapezoidal regions. Let us also point out that the method can be applied whenever the region can be decomposed into regular regions.

BIBLIOGRAPHY AND COMMENTS

<u>Section 1</u>. These methods are discussed in:

Bellman, R. and E. Angel, <u>Dynamic Programming and Partial Differential Equations</u>, Academic Press, Inc., 1972, New York.

There the problems are also treated by means of invariant imbedding. Nonlinear equations can be treated by the method of successive approximations.

Chapter IX

THE THREE DIMENSIONAL POTENTIAL EQUATION

1. INTRODUCTION

In the preceding chapter, we discussed various applications of the theory of dynamic programming to the analytic and computational treatment of the two-dimensional potential equation. A drawback, which becomes major in three dimensions, is the high dimension of the associated Riccati equations. In this chapter we show how this problem can be overcome at the expense of increasing the number of one dimensional operations that are performed. The same method can be applied to many other types of partial differential equations.

2. DISCRETE VARIATIONAL PROBLEMS

Suppose we wish to solve the potential equation

$$u_{xx} + u_{yy} = 0 \tag{1}$$

over the rectangular region $0 \leq x \leq a \quad 0 \leq y \leq b$ where the values of u are specified on the boundary. It is well known that (1) can be replaced by the problem of minimizing the functional

$$J(u) = \int_R (u_x^2 + u_y^2)dR \tag{2}$$

over all functions $u_x, u_y \leq L^2[R]$ and satisfying the desired boundary conditions.

We start then by replacing (2) by an approximate discrete problem. Thus, supposing for simplicity we can find an h such that

$$\begin{aligned}(n+1)h &= a\\ (m+1)h &= b,\end{aligned} \tag{3}$$

where n, m are integers. We seek a solution only for values of x and y given by

$$x = ih \qquad i=1, \ldots, n$$
$$y = jh \qquad j=1, \ldots, m. \tag{4}$$

Defining u_j as

$$u_j = u(ih, jh), \tag{5}$$

we can replace (2) by the approximating discrete functional

$$H(u) = \sum_{i=1}^{n+1} \sum_{j=1}^{m+1} [(u_{ij}-u_{i-1,\ j})^2+(u_{ij}-u_{i,\ j-1})^2]. \tag{6}$$

From the original boundary conditions, we see that the values $\{u_{01}\}$, $\{u_{i0}\}$, $\{u_{n-1,\ j}\}$, $\{u_{i,\ m+1}\}$ are known.

Supppose we now consider the solution of the potential equation, or the corresponding variational problem, over the region of Figure 1 where

$$c = kh,$$
$$d = \ell h \tag{7}$$

k, ℓ integers. Using the discretization of (6) we can approximate this problem by minimizing the discrete functional

$$H_{\ell k}(u) = \sum_{i=1}^{\ell} (u_{k+1,\ j})^2 + \sum_{j=1}^{\ell-1} (u_{k+1,j} - u_{kj})^2 +$$
$$+ \sum_{i=k+2}^{n+1} \sum_{j=1}^{m+1} [(u_{ij}-u_{i-1,\ j})^2+(u_{ij}-u_{i,\ j-1})^2]. \tag{8}$$

It is this second problem we will focus on.

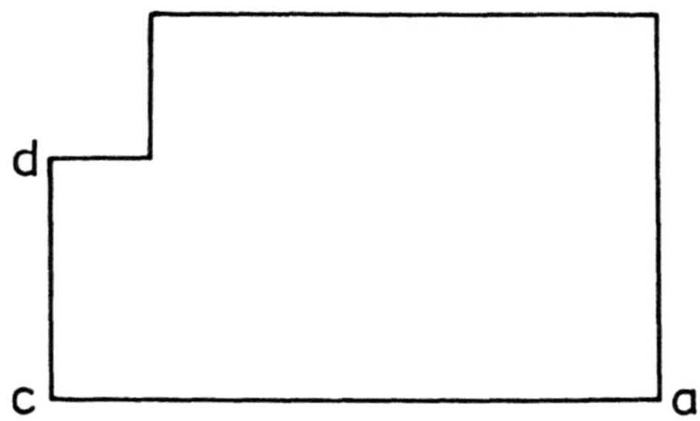

Figure 1

3. DYNAMIC PROGRAMMING

Let us define a vector of values

$$v = (v_1, \ldots, v_m) \tag{1}$$

as follows

$$v_j = u_{kj} \qquad j=1, 2, \ldots, \ell-1$$
$$v_j = u_{k+1,\,j} \qquad j=\ell, \ell+1, \ldots, m. \tag{2}$$

These values are as indicated in Figure 2.

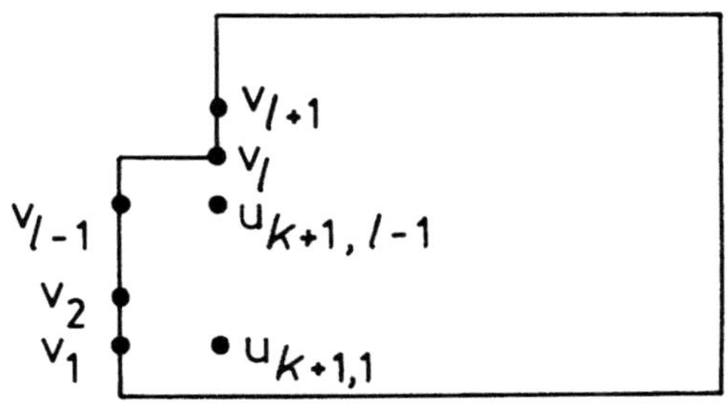

Figure 2

Since the minimum of the discrete functional depends on ℓ, k, and v, let us write

$$f_{\ell k}(v) = \min_{\{u_{ij}\}} H_{\ell k}(u) = \min_{\{u_{ij}\}} \Bigg[\sum_{j=1}^{\ell} (u_{k+1,j} - u_{k+1,\, j-i})^2 +$$
$$+ \sum_{j=1}^{\ell-1} (u_{k+1,j} - u_{kj})^2 + \sum_{i=k+2}^{n+1} \sum_{j=1}^{m+1} (u_{ij} - u_{i-1,\, j})^2 +$$
$$+ (u_{ij} - u_{i,\, j-1})^2 \Bigg] \tag{3}$$

Expanding (3), we have

$$f_{\ell k}(v) = \min_{\{u_{ij}\}} \Big[(u_{k+1,\ell} - u_{k\neq 1,\, \ell-1})^2 + (u_{k+1,\ell-1} - u_{k,\, \ell-1})^2 +$$
$$+ \sum_{j=1}^{\ell-1} (u_{k+1,j} - u_{k+1,\, j-i})^2 + \sum_{j=1}^{\ell-2} (u_{k+1,j} - u_{k,j})^2 +$$

$$+ \sum_{i=k+2}^{n+1} \sum_{j=1}^{m+1} \left[(u_{ij} - u_{i-1,\, j})^2 + (u_{ij} - u_{i,\, j-1})^2 \right] \tag{4}$$

or by (2)

$$f_{\ell k}(v) = \min_{\{u_{ij}\}} \Bigg[(v_\ell - u_{k+1,\, \ell-1})^2 + (u_{k+1,\, \ell-1} - v_{\ell-1})^2 +$$

$$+ \sum_{j=1}^{\ell-1} (u_{k+1,j} - u_{k+1,\, j-1})^2 + \sum_{j=1}^{\ell-2} (u_{k+1,\, j} - u_{k,\, j})^2 +$$

$$- \sum_{i=k+2}^{n+1} \sum_{j=1}^{m+1} \left[(u_{ij} - u_{i-1,\, j})^2 + (u_{ij} - u_{i,\, j-1})^2 \right] \Bigg]. \tag{5}$$

Since the first two terms of this expression depend only on $u_{k+1,\, \ell-1}$ and boundary values, we can move the minimization over the values of u_{ij} other than $u_{k+1,\, \ell-1}$ past these two terms, yielding the equation

$$f_{\ell k}(v) = \min_{u_{k+1,\, \ell-1}} \Bigg[(v_\ell - u_{k+1,\, \ell-1})^2 + (u_{k+1,\, \ell-1} - v_\ell) +$$

$$+ \min_{\{u_{ij} - u_{k+1,\, \ell-1}\}} \Bigg[\sum_{j=1}^{\ell-1} (u_{k+1,j} - u_{k+1,\, j-1})^2 +$$

$$+ \sum_{j=1}^{\ell-2} (u_{k+1,\, j} - u_{k,\, j})^2 + \sum_{i=k+2}^{n+1} \sum_{j=1}^{m+1} (u_{ij} - u_{i-1,\, j})^2 +$$

$$- (u_{ij} - u_{i,\, j-1})^2 \Bigg] \Bigg] \tag{6}$$

However, in view of the definition of $f_{\ell k}(v)$ in (3), we recognize that (6) can be written as

$$f_{\ell k}(v) = \min_w \Big[(v_\ell - w)^2 + (w - v_{\ell-1})^2 +$$

$$+ f_{\ell-1,\, k}(v_1, \ldots, v_{\ell-2}, w, v_\ell, \ldots, v_m) \Big], \tag{7}$$

where we have set $w = u_{k+1,\, \ell-1}$. It is clear that this relation is a consequence of the principle of optimality.

4. BOUNDARY CONDITIONS

In principle, we can now solve (3.7) for values of ℓ from 2 to m+1. However, we would like to also be able to vary k, and we must also specify $f_{\ell k}(v)$.

For k=n, there are no interior points, hence there is no minimization to be performed. Thus we have

$$v_{on}(v) = \sum_{j=1}^{m} (u_{n+1,v_j} - v_j)^2 \sum_{j=1}^{m+1} (u_{n+1,j} - u_{n+1,\ j-1})^2$$

where the region corresponding to f_{on} is as in Figure 3. When we go from $f_{ok}(v)$ to $f_{1k}(v)$, as in Figure 4, we have one additional point but no interior points, so that we have

$$f_{\ell k}(v) = (v_1 - u_{k+1,\ 0})^2 + f_{0k}(v).$$

Finally, it should be clear that

$$f_{0k}(v) = f_{m+1,\ k+1}(v).$$

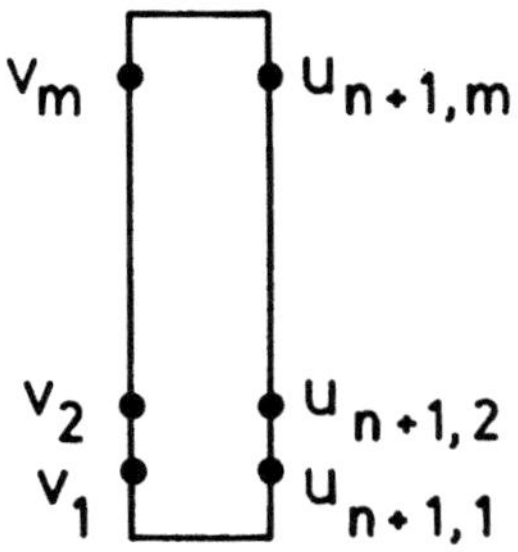

Figure 3

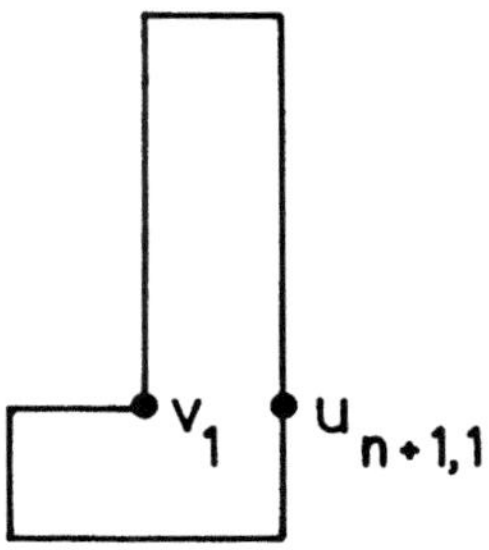

Figure 4

5. RECURRENCE RELATIONS

We can easily prove inductively that $f_{\ell k}(v)$ must be quadratic in v so that we may write

$$f_{\ell k}(v) = \langle A^{(\ell,\ k)}\ v,\ v\rangle + \langle b^{(\ell,\ k)}\ v,\ v\rangle + c^{(\ell,\ k)}, \qquad (1)$$

where, with no loss of generality, we can require $A^{(\ell,\ k)}$ to be symmetric. Using this relation in (3.7) through (4.3), we can obtain recurrence relations for $A^{(\ell,\ k)}$, $b^{(\ell,\ k)}$ and $c^{(\ell,\ k)}$.

Each recurrence relation will involve only scalar operations since only one point is added a: a time.

Suppose now we would like to solve our original equation, the potential equation, over a rectangular region. We must determine $f_{m+1, j}(v)$. We start with $f_{on}(v)$ from (4.1) and then find successively

$$f_{1n}, f_{2n}, \ldots, f_{m}, f_{m+1, n} = f_{o, n-1}, f_{1, n-1},$$

$$\ldots, f_{m+1, n-1} = f_{o, n-2}, \ldots$$

$$\ldots, f_{m+1, 2} = f_{o1}, f_{1, 1}, \ldots, f_{m+1, 1}.$$

For each problem, we must find a new $A^{(\ell, k)}$, $b^{(\ell, k)}$ and $c^{(\ell, k)}$ which must be stored. Once $f_{m+1, 1}$ is determined, we will replace v by the given boundary conditions and proceed backwards determining the minimizing $\{u_{ij}\}$.

6. GENERAL REGIONS

The idea of removing a single point at a time applies to arbitrary regions in n dimensions. For instance, let us consider a three-dimensional region. Then, by removing one mesh point at a time, we will have regions such as in Figure 5. If we define $f_{i\ell k}(v)$ as the minimum of the discrete functional over this region, we can then use dynamic programming to relate $f_{i\ell k}(v)$ to $f_{i\ell k+1}(v)$.

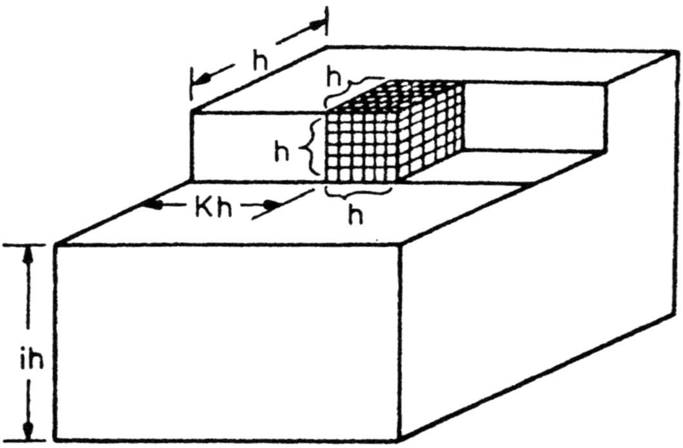

Figure 5

7. DISCUSSION

Proofs of existence, uniqueness and stability follow directly from positive definiteness of the functionals. It is important to note that although we have used the potential equation as an example, we have made use of none of its special properties. Thus, we expect our method to work at least for a large class of elliptic equations. This is in contrast to a number of highly efficient direct methods which usually require the equation to have constant coefficients.

The reader can verify by deriving the recurrence relations for $A^{(\ell, k)}$ and $b^{(\ell, k)}$ that the number of computations which must be performed is fairly high. However, the importance of the method is that it allows one to trade time for storage and thus avoids the dimensionality problem associated with partial differential equations.

Finally, we note that the extensions of these ideas are immediate and far-reaching. Since all we have really required is that the solution at a given point depends only on the solution at adjoining points, we can use the method to reduce the dimensionality difficulties of many kinds of large systems. We can also use invariant imbedding to treat linear equations not necessarily connected with variational questions.

BIBLIOGRAPHY AND COMMENTS

Section 1. We are following the paper:

E. Angel and R. Bellman, 'Reduction of Dimensionality for the Potential Equation using Dynamic Programming', Unititas Mathematica 1 (1972) 181-190.

Chapter X

THE HEAT EQUATION

1. INTRODUCTION

In this chapter, we wish to consider the heat equation.

First, we use the Laplace transform to convert the equation into an equation of potential type. Then we use a numerical inversion of the Laplace transform to obtain the solution of the original equation.

In Sections 2-4, we consider the one-dimensional case where ordinary integration can be used. Some numerical results are given to show the effectiveness of the method.

Finally, in Section 5, we consider the multidimensional case.

The heat equation may be considered a limiting form of a branching process as was pointed out by Feller. It may also be considered a limiting form of a transfer equation where the velocity of the particle is infinite. This allows use of invariant imbedding techniques.

2. THE ONE-DIMENSIONAL HEAT EQUATION

Let us now turn our attention to a partial differential equation where the merit of the Laplace transform resides in converting it into an ordinary differential equation. The equation is

$$k(x)u_t = u_{xx}, \tag{1}$$

subject to the initial boundary conditions

$$\begin{aligned} &u(x, 0) = g(x), \quad 0 < x < 1 \\ &u(0, t) = h_1(t), \quad u(1, t) = h_2(t), \quad t < 0. \end{aligned} \tag{2}$$

This equation may be considered to arise in the following fashion. Take a slender rod of cross-section area Λ and of length L. Let its mass per unit length be ρ, its specific heat be c, and its conductivity be k_1. We denote the temperature at a point x from the left end of the rod at time t by u(x, t), and we symbolize the amount of heat per unit cross-section area per unit time passing x to the right at time t by F(x, t). We focus our attention on the cylinder of base area A extending from x to x+Δ. On the one hand, the rate at which heat is being supplied to it is $c\,\rho\,\Delta\,Au_t$. On the other hand, it is $A[F(x, t)-F(x+\Delta, t)]$. Thus,

$$c\,\rho\,\Delta\,Au_t = -AF_x\Delta, \tag{3}$$

where we have dropped all terms involving powers of Δ higher than the first. In addition, a basic heat conduction postulate is

$$F = -k_1u_x. \tag{4}$$

These equations lead to the desired heat equation,

$$k(x)u_t = u_{xx},$$

where we have set

$$k(x) = \rho c/k_1.$$

We have written k(x) to emphasize that the rod may be inhomogeneous with its density a function of position.

Equations of this type also arise in the study of random walk and diffusion processes.

We shall suppose that $\infty \geq \delta_2 \geq k(x) \geq \delta_1 \geq 0$ for $0 \leq x \leq 1$. Although much weaker conditions are sufficient, this simple condition is appropriate here.

If k(x) is a function of general form, this equation cannot be solved explicitly in terms of the elementary functions of analysis. If numerical results are desired, recourse must be had to computational algorithms. Of these, difference methods are most prominent at the moment. We shall pursue an entirely different approach based upon the use of the Laplace transform.

We have neither an explicit expression for the Laplace transform, nor do we have an initial value problem to solve. Here, as we shall see in a moment in our use of the Laplace transform, we face a two-point, boundary-value problem, fortunately with an associated linear ordinary differential equation.

The technique, in principle, is applicable to high-order parabolic equations such as

$$k(x, y)u_t = u_{xx} + u_{yy}.$$

However, since the calculations for equations of this type can consume an appreciable time, we beg the readers indulgence and ask him to follow our precept rather than our example.

3. THE TRANSFORM EQUATION

Applying the Laplace transform to (2.1), we obtain the equation

$$v\text{-}sk(x)v = -k(x)g(x), \quad 0 < x < 1, \tag{1}$$

with the boundary conditions

$$v(0) = H_1(s), \quad v(1) = H_2(s), \tag{2}$$

where

$$v \equiv v(x, s) = L(u),$$

$$H_1(s) = L(h_1), \qquad H_2(s) = L(h_2). \tag{3}$$

In (1), the prime denotes differentiation with respect to x.

The two-point, boundary-value problem is solved numerically for each value of s we use, in the usual fashion which we now describe. Let $v_1(x)$ and $v_2(x)$ (suppressing the explicit dependence on s) be determined respectively as the solutions of

$$\begin{aligned} v_1'' - sk(x)v_1 &= k(x)g(x), & v_1(0) &= 1, & v_1''(0) &= 0 \\ v_2'' - sk(x)v_2 &= 0, & v_2(0) &= 0, & v_2''(0) &= 1. \end{aligned} \tag{4}$$

There are so-called principal solutions. Returning to (1), set

$$v = H_1(s)v_1 + av_2, \tag{5}$$

where a is a constant to be determined using the second condition in (2). Then, the condition at x = 1 yields

$$H_2(s) = H_1(s)v_1(1) + av_2(1). \tag{6}$$

The condition imposed upon $k(x)$ ensures that $v_2(1) \neq 0$. We ask the reader to accept this fact. These matters are neither simple nor elementary. Thus, we have for a the value

$$a = \frac{H_2(s) - H_1(s)v_1(1)}{v_2(1)}, \tag{7}$$

which via (5) determines the function $v(x)$.

4. SOME NUMERICAL RESULTS

The method was tested for the cases

$$k(x) = 4,\ 9,\ 16, \tag{1}$$

with the initial conditions

$$u(x,\ 0) = \sin \pi x. \tag{2}$$

These were chosen so that we could obtain simple explicit solutions to test the accuracy of our procedure. If $k(x) = k^2$, a constant, the solution of (2.1) is given by

$$u(x,\ t) = e^{-\pi^2 t/k^2} \sin \pi x, \tag{3}$$

as is immediately verified. We used a seven-point quadrature formula and obtained the results indicated in Figures 1, 2 and 3. The larger the value of k, the smaller the conductivity, and the slower the convergence to zero.

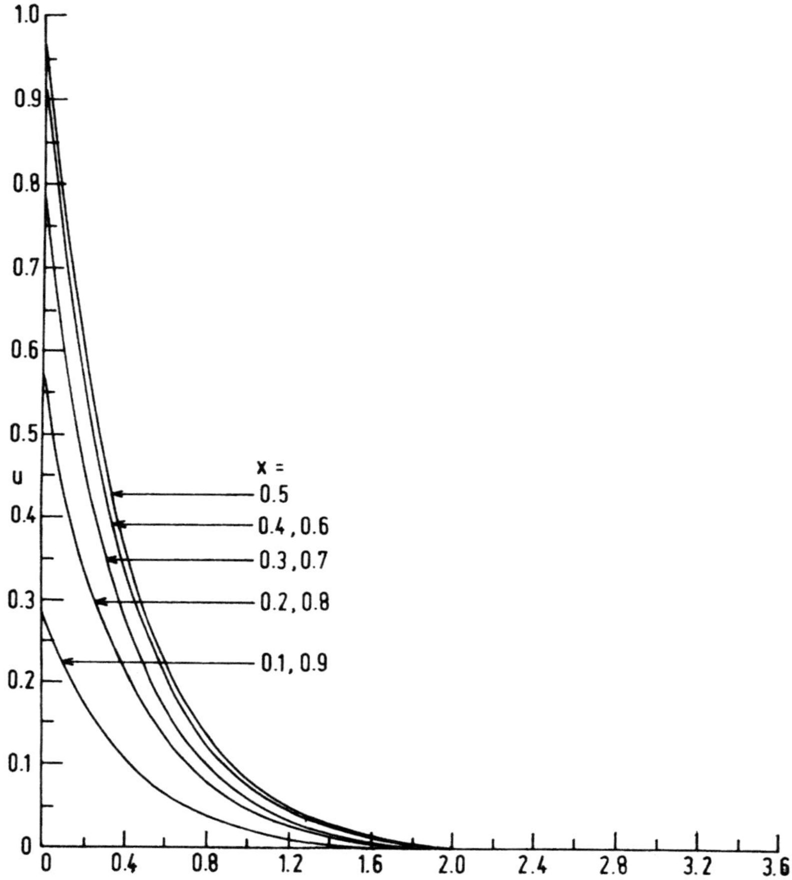

Figure 1. Computational solution for $k(x) \equiv 4$, $g(x) = \sin \pi x$, using a seven-point quadrature formula.

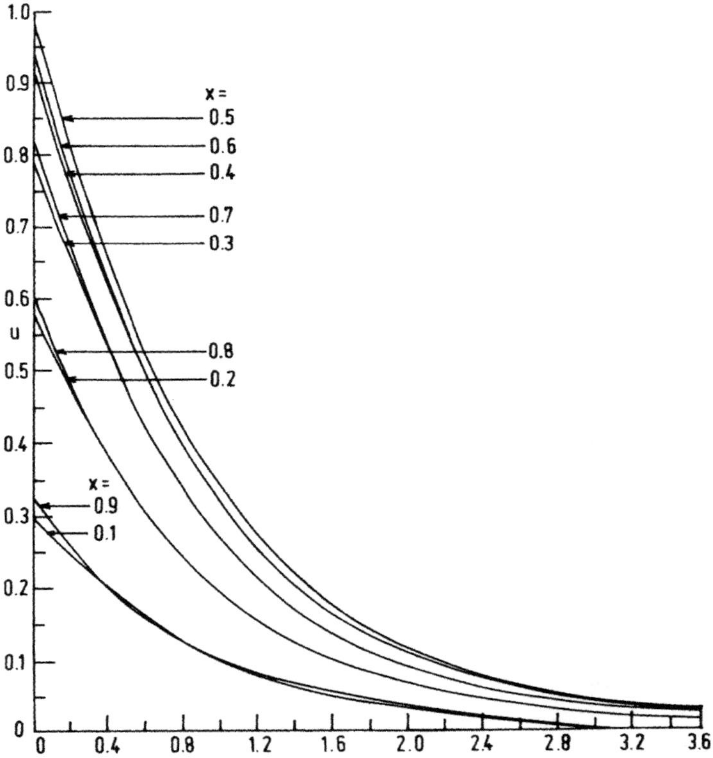

Figure 2. Computational solution for $k(x) \equiv 9$, $g(x) = \sin \pi x$, using a seven-point quadrature formula.

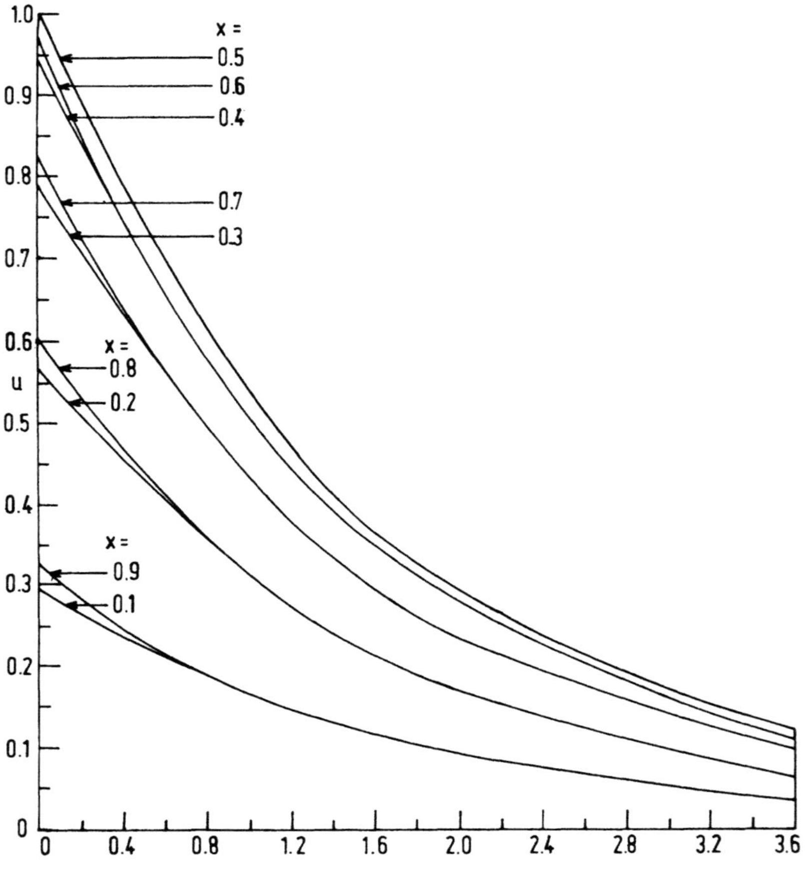

Figure 3. Computational solution for $k(x) \equiv 16$, $g(x) = \sin \pi x$, using a seven-point quadrature formula.

Figures 4 and 5 show the results for an inhomogeneous rod for which

$$k(x) = 1+x. \tag{4}$$

Here we have no simple explicit solution. The calculations were done with both seven- and nine-point quadrature formulas in order to obtain a self-consistent check on the accuracy.

The production of the data for one of the graphs took about thirty seconds on an IBM-7090. In the integration of (3.4) and (3.5), a Runge-Kutta procedure with a grid size of 0.01 was used.

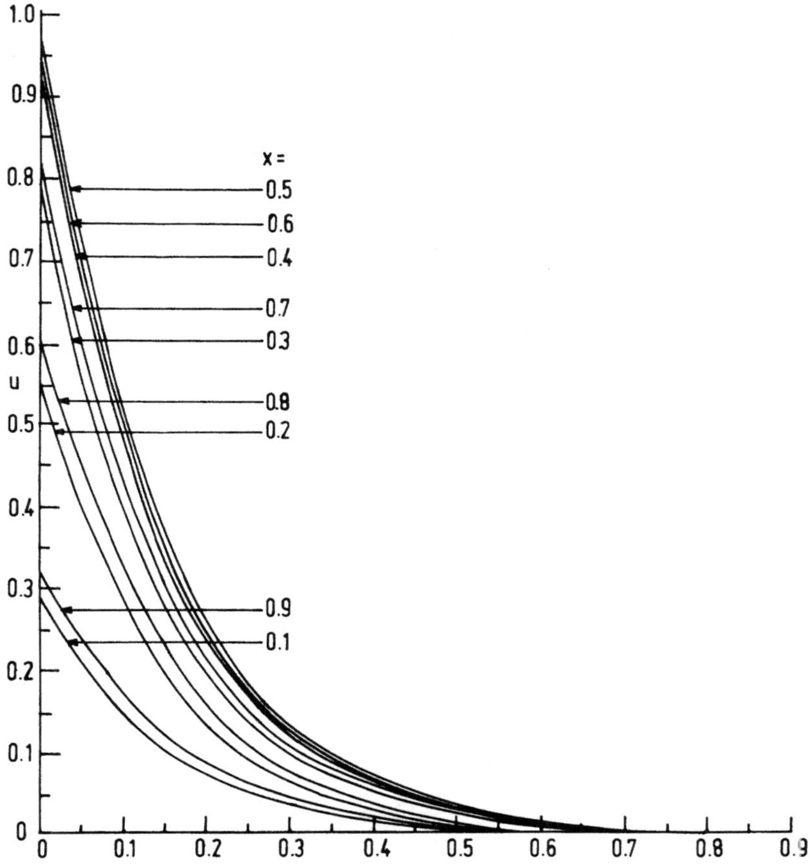

Figure 4. Computational solution for $k(x) \equiv 1 + x$, $g(x) = \sin \pi x$, using a seven-point quadrature formula.

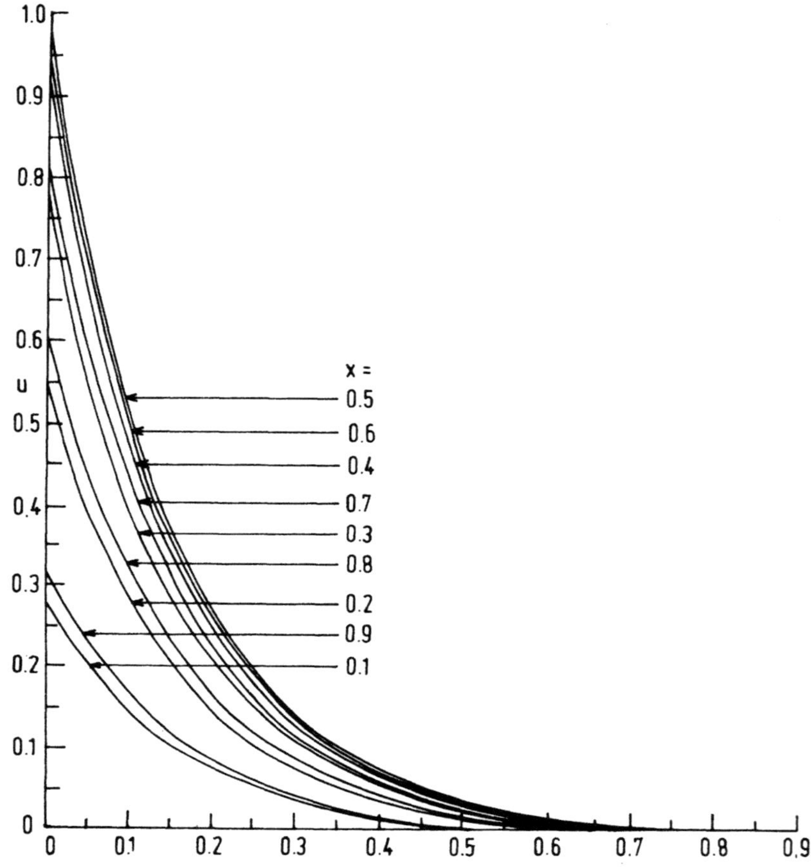

Figure 5. Computational solution for $k(x) \equiv 1 + x$, $g(x) = \sin \pi x$, using a nine-point quadrature formula.

5. MULTIDIMENSIONAL CASE

In the multidimensional case, after we take the Laplace transform, we are left with an equation of potential type. We can therefore use the methods of Chapters 8 and 9 to solve this equation. After we have obtained the solution, we use a numerical inversion technique to find a solution of the original heat equation.

Let us note that the use of a numerical inversion technique automatically furnishes selective computation.

BIBLIOGRAPHY AND COMMENTS

Section 1. Numerical inversion methods for the Laplace transform are given in:

Bellman, R., Introduction to Matrix Analysis, McGraw-Hill Book Co., New York, 1970.
Bellman, R. and R. Roth, Laplace Transform, to appear.

For a direct use of invariant imbedding techniques on the heat equation, see:

Bellman, R. and G.M. Wing, 'Hydrodynamical Stability and Poincaré-Lyapunov Theory', Proc. of the National Academy of Sciences, Vol. 42, 867-870, 1956.

Another procedure is to use the Laplace transform to obtain a steady state equation. Invariant imbedding may be applied to this equation. Once the invariant imbedding equation is obtained, we may then invert. This avoids the use of particle counting techniques. For the use of this technique see the book on the Laplace transform cited above.

For the use of invariant imbedding to study the wave equation, see:

Bellman, R. and R. Vasudevan, Wave Propagation - An Imbedding Approach, to be published by D. Reidel, Dordrecht (1984).

Section 5. Selective Computation is discussed in:

Bellman, R., Selective Computation, to appear.

Chapter XI

NONLINEAR PARABOLIC EQUATIONS

1. INTRODUCTION

In this chapter we wish to consider the nonlinear parabolic equation

$$u_t - u_{xx} = g(u)$$
$$u(0, t) = u(1, t) = 0 \qquad (1)$$
$$\lim_{t \to 0} u(x, t) = f(x)$$

We wish to establish an analogue of the classical Poincaré-Lyapunov theorem:

If g(u) is a power series in u lacking constant and first degree terms if $\|f\|$ is sufficiently small, then the solution exists for all positive t.

Here we use the norm

$$\|f\| = \max_x |f|. \qquad (2)$$

In Sections 2-5 we present some results for the linear equation of interest in themselves. These results are crucial for our treatment of the nonlinear case. In Section 7, we consider the nonlinear equation. In Section 8 we say a few words about asymptotic behavior. In Section 9, we indicate some extensions.

In Chapter 1, we use the positivity of the linear equation to treat the nonlinear equation in a different fashion.

2. LINEAR EQUATION

Let us now consider the linear equation. As we know the solution may be written

$$u = \int_0^1 k(x, y, t)f(y)dy. \qquad (1)$$

The kernel, $k(x, y, t)$, is an analogue of the matrix exponential. As for the matrix exponential, a simple semigroup argument yields the identity

$$k(x, y, s+t) = \int_0^1 k(x, z, t)k(z, y, s)dz. \tag{2}$$

This was pointed out by Hadamard. Use of this relation yields another method for the calculation of the solution of the linear equation.

In Chapter 10, we gave a derivation of the linear equation. It will be clear from this derivation how inhomogeneity can enter. Here and below, we assume that this inhomogeneity is uniformly positive.

Let us make some observations about the linear equation. The same equation governs a diffusion process. This equation may be considered a limiting form of an equation which occurs in the study of branching processes, as was pointed out by Feller. The equation may also be taken to be a limiting form of a transport equation where the particle velocity is infinite. This allows the use of invariant imbedding techniques.

Let us now consider the inhomogeneous equation

$$u_t - u_{xx} = h. \tag{3}$$

The solution may be written

$$u = v + \int_0^t \int_0^t k(x, y, t_s)h \, dy \, ds. \tag{4}$$

3. THE NON-NEGATIVITY OF THE KERNEL

We shall study the kernel by various techniques.

Let us note that one technique we can employ is to use a difference approximation, as we did in Chapter 1. The case where the equation is inhomogenous leads to a simple modification the non-negativity of the solution is apparent whenever the initial value is non-negative.

4. MONOTONICITY OF MEAN VALUES

The result that follows was used by Bochner to treat more general problems. As we shall show, it can readily be used to establish non-negativity. Let us begin with the simplest case.

If

$$u_t = u_{xx}, \qquad 0 < x < 1, \quad t > 0,$$
$$u(0, t) = u(1, t) = 0, \qquad t > 0, \tag{1}$$
$$u(x, 0) = g(x), \qquad 0 < x < 1,$$

then

$$I_n(t) = \int_0^1 [u(x, t)]^{2n} \, dx, \qquad n=1, 2, \ldots, \tag{2}$$

is a monotone decreasing function of t for $t \geq 0$. From this it follows that

$$\max_{0 \leq x \leq 1} |u(x, t)| \tag{3}$$

is a monotone decreasing function of t for $t \geq 0$. The result can easily be extended to cover general types of parabolic equations. We have

$$\begin{aligned} \frac{dI_n}{dt} &= 2n \int_0^1 [u(x, t)]^{2n-1} u_t \, dx \\ &= 2n \int_0^1 [u(x, t)]^{n-1} u_{xx} \, dx \\ &= 2n\{[u(x, t)]^{2n-1} u_x\}_0^1 - 2n(2n-1) \int_0^1 [u(x, t)]^{2n-2} u_x^2 dx \\ &= -2n(2n-1) \int_0^1 u^{2n-2} u_x^2 \, dx. \end{aligned} \tag{4}$$

Hence, $dI_n/dt < 0$.

To show that the function (3) has the desired property, we use the following result of interest in itself.

If $v(x)$ is a continuous function in a finite interval $[a, b]$, then

$$\lim_{n\to\infty}\left\{\int_a^b [v(x)]^{2n}\,dx\right\}^{\frac{1}{2}n} = \max_{a\le x\le b} |\, v(x)\,|. \tag{5}$$

The proof is easily obtained. Let $m = \max |\, v(x)\,|$ in $[a, b]$; then

$$m^{2n}(b-a) \ge \int_a^b [v(x)]^{2n}\,dx \ge \int_c^d [v(x)]^{2n}\,dx, \tag{6}$$

where $[c, d]$ is a subinterval of $[a, b]$ within which $|\, v(x)\,| \ge m-\varepsilon$. Thus

$$m(b-a)^{\frac{1}{2}n} \ge \left\{\int_a^b [v(x)]^{2n}\,dx\right\}^{\frac{1}{2}n} \ge (m-\varepsilon)(d-c)^{\frac{1}{2}n} \tag{7}$$

Since

$$\lim_{n\to\infty} q^{\frac{1}{2}n} = 1, \tag{8}$$

for any $q > 0$, we see that (5) holds.

Combining the two results, we have a proof of the theorem. The preceding result appears first to have been used by M. Riesz.

It is easy to see how we modify the integral to handle the inhomogenous case. In the multidimensional case, we use Green's theorem in place of integration by parts.

5. POSITIVITY OF THE PARABOLIC OPERATOR

Let us now, following Bochner, use the monotone behavior of the function

$$\max_{0\le x\le 1} |\, u(x, t)\,|,$$

established above, to show the positivity of the parabolic operator with appropriate boundary conditions.

The solution of

$$u_t = u_{xx}, \qquad u(x, 0) = f(x), \tag{1}$$

$t > 0$, $f(x)$ periodic in x of period 2π, is given by

$$u(x, t) = \frac{1}{2\pi} \int_0^{2\pi} G(x-y, t) f(y)\, dy, \tag{2}$$

where

$$G(x, t) = 1 + 2 \sum_{\gamma=1}^{\infty} e^{-\gamma^2 t} \cos \gamma x. \tag{3}$$

The positivity of $G(x, t)$ follows directly from the functional equation for the theta function, namely, the fundamental transformation formula

$$G(x, t) = \left(\frac{\pi}{t}\right)^{\frac{1}{2}} \sum_{\gamma=-\infty}^{\infty} \exp \frac{-(x-2\pi\gamma)^2}{4t} \tag{4}$$

Relations of the type (4), however, are not available for general regions, whereas the following argument is independent of the region.

Suppose that $G(x, t)$ were negative for some $t > 0$ and some x. If this value of t is kept fixed, the normalization condition

$$\frac{1}{2\pi} \int_0^{2\pi} G(y, t)\, dy = 1 \tag{5}$$

implies that there must be an open set R such that

$$\frac{1}{2\pi} \int_R G(y, t)\, dy > 1. \tag{6}$$

Hence, there is an interval $I \subset R$ such that

$$\frac{1}{2\pi} \int_I G(y, t)\, dy > 1. \tag{7}$$

We now proceed along classic lines in the theory of partial differential equations. It is possible to construct a continuous smoothing function f(x) such that

$$\begin{aligned} &(a)\ 0 \le f(x) \le 1 \quad \text{for all } x, \\ &(b)\ f(x) = 1, \quad x \in I, \\ &(c)\ f(x) = 0 \quad \text{outside } R. \end{aligned} \tag{8}$$

For this function, and for all x and t, we have

$$| u(x, t) | \le \max_x | f(x) | = 1; \tag{9}$$

but for the value of t for which G(x, t) is assumed to be negative, we have

$$\begin{aligned} u(0, t) &= \frac{1}{2\pi} \int_0^{2\pi} G(y, t) f(y)\, dt \\ &= \frac{1}{2\pi} \int_I G(y, t)\, dy > 1, \end{aligned} \tag{10}$$

a contradiction.

6. NONLINEAR EQUATIONS

Let us now consider the nonlinear equation. This is done in the following steps:

(1) Conversion of the nonlinear partial differential equation into a nonlinear integral equation.

(2) Use of successive approximation to solve the nonlinear integral equation.

(3) Uniform boundedness of the successive approximations in terms of $\|f\|$.

(4) Convergence of the successive approximations.

Using the result above, we have the nonlinear integral equation

$$u = v + \int_0^1 \int_0^T k(x, y, t-s)\, g(u)\, dy\, ds. \tag{1}$$

This is treated by means of successive approximations. We obtain these from the relation

$$u_{n+1} = v + \int_0^1 \int_0^T k(x, y, t-s)\, g\,(u_n)\, dy\, ds. \tag{2}$$

By virtue of the assumption concerning g we have the uniform bound

$$\|u_n\| \leq 2\|f\|. \tag{3}$$

Also from the assumption concerning g we see that we have geometric convergence of the series whose general terms is

$$\|u_{n+1} - u_n\| \tag{4}$$

We have merely sketched the proof since the steps are the same as for ordinary differential equations. The essential results are given in Sections 2-4.

7. ASYMPTOTIC BEHAVIOR

It is easy to see that we have

$$u \sim T(f)e^{-\pi t}. \tag{1}$$

The coefficient T(f) may be obtained by iteration in the nonlinear integral equation. In the case of ordinary differential equations, we can use invariant imbedding to obtain an equation for this coefficient. This does not seem possible here.

8. EXTENSIONS

Let us now indicate some possible extensions of the foregoing result.

We can, first of all, consider the inhomogenous case. Secondly, we can allow many different boundary conditions. Thirdly, we can consider the multidimensional case. Here, we have to employ some results from the theory of integral equations. Finally, a similar result can be obtained for many different classes of equations.

BIBLIOGRAPHY AND COMMENTS

Section 1. This result was given in:

Bellman, R., 'On the Existence and Boundedness of Solutions of Nonlinear Partial Differential Equations of Parabolic Types'. Transaction of the American Mathematical Society 64 (1948) 21-44.

The result for ordinary differential equations may be found in:

Bellman, R., Methods of Nonlinear Analysis, Vol. I, Academic Press, Inc., New York, 1969.

There is an analogous result for differential difference equations. It may be found in:

Bellman, R. and K.L. Cooke, Differential-Difference Equations, Academic Press, Inc., New York, 1963.

See also:

Bellman, R., 'Equilibrium Analysis, the Stability Theory of Poincaré and Lyapunov and Extensions', Modern Mathematics for the Engineer (1956) 30-35.

Bellman, R. and G.M. Wing, 'Hydrodynamical Stability and Poincaré-Lyapunov Theory', Proceedings of the National Academy of Sciences 42 (1956) 867-870.

Bellman, R., 'On Analogues of Poincaré-Lyapunov Theory for Multipoint Boundary-Value Problems - I', Journal of Mathematical Analysis and Applications 14 (1966) 522-526.

Bellman, R., 'On Analogues of Poincaré-Lyapunov Theory for Multipoint Boundary-Value Problems - Correction', Journal of Mathematical Analysis and Applications 14 (1966) 522-526.

Section 2. The result for the matrix exponential may be found in:

Bellman, R., Selective Computation, to appear.

Section 4. See:

Beckenbach, E.F. and R. Bellman, Inequalities, Springer-Verlag. Berlin, 1970.

<u>Section 6</u>. The result for ordinary differential equations may be found in the book cited above.

<u>Section 7</u>. The use of invariant imbedding for ordinary differential equations is given in the book on selective computation cited above.

Chapter XII

DIFFERENTIAL QUADRATURE

1. INTRODUCTION

The numerical solution of nonlinear partial differential equations plays a prominent role in numerical weather forecasting, optimal control theory, radiative transfer, biology, and many other areas of physics, and engineering. In many cases all that is desired is a moderately accurate solution at a few points which can be calculated rapidly. The standard finite difference methods currently in use have the characteristic that the solution must be calculated at a large number of points in order to obtain moderately accurate results at the points of interest. Consequently, both the computing time and storage required often prohibit the calculation. Furthermore, the mathematical techniques involved in finite difference schemes or in the Fourier transform methods, are often quite sophisticated and thus not easily learned or used.

In this chapter we wish to present a technique which can be applied in a large number of cases to circumvent both the above difficulties. We illustrate this technique with the solution of some partial differential equations arising in various simplified models of fluid flow, turbulence, and biology.

2. DIFFERENTIAL QUADRATURE

Consider the nonlinear first-order partial differential equation

$$u_t = g(t, x, u, u_z), \quad -\infty < x < \infty, \quad t > 0 \tag{1}$$

with initial condition

$$u(0, x) = h(x), \tag{2}$$

an equation arising in many mathematical models of physical processes. Let us make the assumption that the function u satisfying Equations (1) and (2) is sufficiently smooth to allow us to write the approximate relation

$$u_x(t, x_i) \cong \sum_{j=1}^{N} a_{ij} u(t, x_j), \quad i=1, 2, \ldots, N. \tag{3}$$

There are many ways of determining the coefficients a_{ij}. One method for determining these coefficients will be discussed below. Substitution of Equation (3) into Equation (1) yields the set of N ordinary differential equations

$$u_t(t, x_i) \cong g\left(t, x_i, u(t, x_i), \sum_{j=1}^{N} a_{ij} u(t, x_j)\right) \tag{4}$$

with initial conditions

$$u(0, x_i) = h(x_i), \quad i=1, 2, \ldots, N. \tag{5}$$

Hence, under the assumption that (3) is valid, we have succeeded in reducing the task of solving Equation (1) to the task of solving a set of ordinary differential equations with prescribed initial values.

The numerical solution of such a system, Equation (5), is a simple task for modern electronic computers using standard programs with minimal computing time and storage (in contrast to the case of partial differential equations). The additional storage needed for the solution of a partial differential equation using the differential quadrature technique is that of storing the $N \times N$ matrix a_{ij}.

In practice, relatively low order differential quadrature is sufficient; consequently the total amount of storage and time required on the machine is quite low. We also note that the number of arithmetic operations to be performed at each point are the N additions and multiplications, Equation (3) plus the amount needed for the solution of the set of ordinary differential equations (which is, of course, method-dependent).

3. DETERMINATION OF WEIGHTING COEFFICIENTS

In order to determine appropriate coefficients in the approximation

$$u_x(x_i) \cong \sum_{j=1}^{N} a_{ij} u(x_j), \quad i=1, 2, \ldots, N, \tag{1}$$

we can proceed by analogy with the classical quadrature case, demanding that Equation (1) be exact for all polynomials of degree less than or equal to N-1. The test function $p_k(X) = X^{k-1}$, k=1, ..., N, for arbitrary distinct X_i leads to the set of linear algebraic equations

$$\sum_{j=1}^{N} a_{ij} X_j^{k-1} = (k-1) X_i^{k-1}, \quad k=1, 2, \ldots, N. \tag{2}$$

which has a unique solution since the coefficient matrix is a Vandermonde matrix.

Rather than solving a set of linear algebraic equations, we may readily determine the a_{ij} explicitly once and for all if X_i are properly selected. In this discussion the X_i are chosen to be the roots of the shifted Legendre polynomial of degree N, $P_N^*(X)$, with orthogonality range of [0, 1]. The polynomials $P_N^*(X)$ are defined in terms of the usual Legendre polynomials by the relation

$$P_N^*(X) = P_N(1-2x). \tag{3}$$

The nodes X_i of $P_N^*(X)$, N = 7,9 are given in Table 1.

TABLE I
Nodes of $P_N^*(X)$

X	N = 7	N = 9
X_1	0.2544604	0.01591988
X_2	0.12923440	0.08198445
X_3	0.29707742	0.19331428
X_4	0.50	0.33787329
X_5	0.70292257	0.50
X_6	0.80076559	0.66212671
X_7	0.97455395	0.80668572
X_8	-	0.91801555
X_9	-	0.98408012

By analogy with Lagrange's interpolation formula the test function is taken to be of the form

$$p_k(X) = P_N^*(X)/[(X - X_k)P_N^{*\prime}(X_k)]. \quad (4)$$

It follows that $p_k(X)$ is a polynomial of degree $(N-1)$ such that $p_k(X_j) = \delta_{kj}$. Using the fact that Equation (1) is to be exact for $u(X) = p_k(X)$, we see that

$$a_{ik} = P_N^{\prime *}(X_i)/[(X_i - X_k)P_N^{\prime *}(X_k)], \quad i \neq k. \quad (5)$$

For the case k=i, use of L'Hospital's rule plus the fact that $P_N^*(X)$ satisfies the differential equation

$$X(1-X^2)P_N^{*\prime\prime}(X) + (1+2X)P_N^{*\prime}(X) + N(N+1)P_N^*(X) = 0 \quad (6)$$

given

$$a_{kk} = (1-2X_k)/[2X_k(X_k-1)]. \quad (7)$$

Using Equations (3), (5) and (7), the constant coefficients a_{ij} for any value of N, say N=3, ..., 15, can be easily calculated. These values are then used as input data for a given problem once N has been chosen.

4. NUMERICAL RESULTS FOR FIRST ORDER PROBLEMS

A number of computational experiments were carried out to test the efficacy of the above approach. In all numerical examples given, the reduced set of ordinary differential equations was solved by an Adams-Moulton integration scheme with step size of 0.01. The limits of integration (unless specified) were from t=0 to t=1.

The first numerical experiment was carried out for the equation

$$u_t(X, t) = X^2 - \frac{1}{4}u_x^2(X, t), \quad (1)$$

$$u(X, 0) = 0, \quad (2)$$

which arises in the theory of dynamic programming. Replacing the derivative term on the right side of Equation (1) by an approximating sum, we obtain the nonlinear set of ordinary differential equations

$$u_t(X_i, t) = X_i^2 - \frac{1}{4}\left[\sum_{j=1}^{N} a_{ij}\dot{u}(X_j, t)\right]^2 \tag{3}$$

with initial conditions

$$u(X_i, 0) = 0, \quad i=1, 2, \ldots, N. \tag{4}$$

The analytic solution of Equation (1) is known to be

$$u(X, t) = X^2 \tanh(t), \tag{5}$$

which gives a means of checking our numerical results. For a quadrature of order N = 7, results are displayed under the column e(1) of Table II. The quantity e(k) is defined to be the relative error in computation, i.e., the ratio of the absolute error to the absolute value of the actual analytical answer. Index k = 1, 2, 3 and 4 refers to the different experiments consider in this section.

TABLE II
Differential Quadrature of Order N = 7

t	X	e(1)	e(2)	e(3)	e(4)
0.1	X_1	6. e-08	2. e-10	2. e-05	8. e-08
0.1	X_4	8. e-08	3. e-10	6. e-06	1. e-10
0.1	X_7	1. e-07	1. e-07	3. e-05	2. e-07
0.5	X_1	4. e-06	7. e-08	6. e-04	5. e-07
0.5	X_4	9. e-08	2. e-08	8. e-05	1. e-07
0.5	X_7	5. e-07	8. e-07	3. e-04	1. e-06
1.0	X_1	3. e-05	1. e-07	1. e-04	1. e-04
1.0	X_4	2. e-07	4. e-08	2. e-04	3. e-05
1.0	X_7	7. e-07	2. e-06	1. e-03	1. e-03

The second set of numerical experiments concerned a hyperbolic nonlinear problem of the form

$$u_t(X, t) = uu_x(X, t), \quad \text{in } 0 < x \leq 1, \quad 0 \leq t \leq T, \tag{6}$$

$$u(X, 0) = g(x), \qquad \text{for } 0 < x \leq 1. \tag{7}$$

This is a well-known test equation which possesses the implicit solution

$$u(X, t) = g(x + ut). \tag{8}$$

The shock phenomenon inherent in the solution of Equation (6) can be pushed far into the future by a suitable selection of g, and thus will not be considered here. As a first example we let $g(x) = 0.1x$. In this case the exact solution of Equation (6) is

$$u(X, t) = X/(t-10). \tag{9}$$

Replacing the X-derivative by a differential quadrature of order $N = 7$ and integrating the resulting set of equations with $T = 1$ we obtain the results of Table II, Column e(2).

In both examples 1 and 2, excellent results were to be expected since the solution of the equation chosen was a polynomial in the discretized variable. The experiments were thus tests of the numerical stability of the algorithm.

We now examine equations where the solution is more complex. Consider Equation (6) with the initial function

$$g(X) = (0.1) \sin \pi X. \tag{10}$$

The analytic solution is

$$u(X, t) = (0.1) \sin \pi (X+ut), \tag{11}$$

with a well-behaved solution for $0 \leq t \leq 1$. We compute the solution of Equation (11) by the Newton-Ralphson method, using as our initial approximation the computed value obtained from the differential quadrature version of Equation (6). The relative error for $N = 7$ is given by e(3) of Table II.

The last experiment in this section involves solving Equation (6) with the initial condition

$$g(X) = 0.2X^2. \tag{12}$$

In this case the explicit analytic solution is

$$u(X, t) = \frac{(1 - (0.4)\, tX) - \sqrt{(1 - (0.8)\, tX)}}{(0.4)t^2} \tag{13}$$

Using the same order quadrature and integration scheme as before, we obtain e(4) of Table II.

5. SYSTEMS OF NONLINEAR PARTIAL DIFFERENTIAL EQUATIONS

Consider next the nonlinear system of equations

$$u_t = uu_x + vu_y, \qquad u(x, y, 0) = f(x, y), \tag{1}$$

$$v_t = uv_x + vv_y, \qquad v(x, y, 0) = g(x, y). \tag{2}$$

We wish to use differential quadrature to obtain numerical values for the functions u and v. To check the accuracy of our results, we note that Equations (1) and (2) possess the implicit solution

$$u(x, y, t) = f(x+tu, y+tv), \tag{3}$$

$$v(x, y, t) = g(x+tu, y+tv), \tag{4}$$

a straightforward extension of the one-dimensional case above. Adopting the notation

$$u_{ij}(t) = u(x_i, y_j, t), \tag{5}$$

$$v_{ij}(t) = v(x_i, y_j, t), \tag{6}$$

and using the foregoing approximations for the partial derivatives, we may write Equations (4) and (5) as the system of ordinary differential equations

$$u'_{ij}(t) = u_{ij}(t) \sum_{k=1}^{N} a_{ik}u_{kj}(t) \sum_{k=1}^{N} a_{jk}u_{ik}(t), \tag{7}$$

$$v'_{ij}(t) = u_{ij}(t) \sum_{k=1}^{N} a_{ik}v_{kj}(t) + v_{ij}(t) \sum_{k=1}^{N} a_{jk}v_{ik}(t), \tag{8}$$

$$i, j=1, 2, \ldots, N.$$

The initial conditions are

$$u_{ij}(0) = f(x_i, y_j), \tag{9}$$

$$v_{ij}(0) = g(x_i, y_j), \qquad i, j=1, 2, \ldots, N. \tag{10}$$

Using the above reduction, numerical experiments were carried out for a number of different initial functions f and g. Since the solution of Equations (1) and (2) can also possess a shock phenomenon for finite t, care was taken in selecting f and g to insure that the shock took place for a value of t far away from our region of interest.

The first experiment involved the case

$$f(x, y) = g(x, y) = x+t. \tag{11}$$

The explicit solution is given by

$$u(x, y) = v(x, y) = (x+y)/(1-2t). \tag{12}$$

Using Equations (7) and (8) with N = 7 and an integration step size of 0.01, the results corresponding to $e_u(1) = e_v(1)$ of Table III were obtained.

TABLE III
Differential Quadrature of Order N = 7

t	x	y	$e_u(1) = e_v(1)$	$e_u(2)$	$e_v(2)$
0.1	x_1	y_1	9. e-08	5. e-07	5. e-09
0.1	x_7	y_2	9. e-08	5. e-07	5. e-09
0.1	x_4	y_4	9. e-08	9. e-08	5. e-09
0.1	x_1	y_7	9. e-08	4. e-07	5. e-09
0.1	x_7	y_7	9. e-08	5. e-07	5. e-09
0.2	x_1	y_1	6. e-07	1. e-04	6. e-10
0.2	x_7	y_1	6. e-07	1. e-03	6. e-10
0.2	x_4	y_4	6. e-07	3. e-05	6. e-10
0.2	x_1	y_7	6. e-07	1. e-04	6. e-10
0.2	x_7	y_7	6. e-07	1. e-03	6. e-10
0.3	x_1	y_1	3. e-06	-	-
0.3	x_7	y_1	3. e-06	-	-
0.3	x_4	y_4	3. e-06	-	-
0.3	x_4	y_7	3. e-06	-	-
0.3	x	y_7	3. e-06	-	-

Next, the initial functions were changed to $f(x, y) = x^2$, $g(x, y) = y$. The solution for this case is

$$u(x, y, t) = [(1-2tx) - \sqrt{1-4tx}]/2t^2 \tag{13}$$

$$v(x, y, t) = y/(1-t). \tag{14}$$

The shock occurs at $t=x/4$. Integrating Equations (7) and (8) from t=0 to t=0.20 with N=7, we obtain $e_u(2)$ and $e_v(2)$ of Table III which are relative errors in functions u and v, respectively.

The final case was for the initial functions

$$f(x, y) = 0.1 \sin \pi x \sin \pi y/2, \tag{15}$$

$$g(x, y) = 0.1 \sin \pi x/2 \sin \pi y. \tag{16}$$

Since no explicit solution exists in this case, Equations (3) and (4) together with the Newton-Raphson method were used to produce the solution. Table IV summarizes the results for the differential quadrature of order N = 7. The values corrsponding to $x_k = y_k$, k=1, ..., N for N = 7 are listed in Table IV.

TABLE IV
Differential Quadrature of Order 7

t	x	y	$e_u(3)$	$e_v(3)$
0.1	x_1	y_1	5. e-07	5. e-07
0.1	x_7	y_1	4. e-06	2. e-05
0.1	x_4	y_4	3. e-06	3. e-06
0.1	x_1	y_7	2. e-05	4. e-06
0.1	x_7	y_7	3. e-05	3. e-05
0.5	x_1	y_1	7. e-06	7. e-06
0.5	x_7	y_1	1. e-04	6. e-04
0.5	x_4	y_4	8. e-05	8. e-05
0.5	x_1	y_7	6. e-04	1. e-04
0.5	x_7	y_7	3. e-04	3. e-04
1.0	x_1	y_1	2. e-05	2. e-05
1.0	x_7	y_1	7. e-04	9. e-05
1.0	x_4	y_4	1. e-04	1. e-04
1.0	x_1	y_7	9. e-05	7. e-04
1.0	x_1	y_7	1. e-03	1. e-03

6. HIGHER ORDER PROBLEMS

We have seen that a good approximation to the first derivative of a function may often be obtained in the form

$$u_x(X_i) = \sum_{j=1}^{N} a_{ij}u(X_j), \quad i=1, 2, \ldots, N. \tag{1}$$

Let us now indicate how this idea may be extended to higher order derivatives.

Viewing Equation (1) as a linear transformation of u,

$$u_x = Au, \tag{2}$$

we see that the second-order derivatives can be approximated by

$$u_{xx} = (u_x)_x = A(Au) = A^2u, \tag{3}$$

Thus higher-order derivative approximations are obtained by iterating the linear transformation A.

When the method is applied to two and higher dimensional systems, the choice of N becomes critical. The previous results and the following numerical experiment shows that low order approximations can be expected to yield both qualitative and quantitative information.

Let us now apply the foregoing ideas to the treatment of Burger's equation

$$u_t + uu_x = \varepsilon u_{xx}, \quad \varepsilon > 0 \tag{4}$$

with pure initial value condition

$$u(0, X) = f(X) \tag{5}$$

Burger's equation enjoys a Riccati-like property in the sense that its solutions are expressible in terms of the solution of the linear heat equation

$$w_t = \varepsilon w_{xx}, \tag{6}$$

$$w(0,X) = g(X), \tag{7}$$

by the transformation

$$u = -2\varepsilon w_x/w, \tag{8}$$

$$f(X) = 2\varepsilon g_x(X)/g(X). \tag{9}$$

This property will allow us to compare our numerical results with analytic solutions of Equations (6)-(9).

Adopting the notation

$$u_i(t) = u(t, x_i) \tag{10}$$

yields the set of ordinary differential equations

$$u_i'(t)+u_i(t)\sum_{j=1}^{N} a_{ij}u_j(t) = \varepsilon \sum_{k=1}^{N}\sum_{j=1}^{N} a_{ik}a_{kj}u_j(t),$$

$$i=1, 2, \ldots, N. \tag{11}$$

The initial conditions at t=0 are determined by the initial function

$$u_i(0) = f(x_i), \quad i=1, 2, \ldots, N. \tag{12}$$

In the following numerical experiment, the initial function f(x) was chosen to be

$$f(x) = -2\varepsilon[b\pi \cos \pi x+c\pi/2 \cos \pi x/2]/[b \sin \pi x+c \sin \pi x/2+d], \tag{13}$$

where b, c, d are constants. The analytic solution of the heat equation in this case is

$$w(t, x) = be^{-\varepsilon\pi^2 t} \sin \pi x+ce^{-\varepsilon\pi^2 t/4} \sin \pi x/2+d \tag{14}$$

Table V summarizes the results of two cases N = 7 and 9 order differential quadrature. The values of the constants are taken to be ε = 0.01, b = 0.2, c = 0.1, and d = 0.3. The relative errors e_u and e_w correspond to the Burgers' and the heat equation, respectively. Values of x_k are those listed in Table I.

TABLE V
Differential Quadrature of Orders 7 and 9

	N = 7			N = 9		
t	x	e_u	e_w	x	e_u	e_w
0.1	x_1	4. e-04	2. e-05	x_1	1. e-05	6. e-07
0.1	x_3	2. e-05	2. e-07	x_3	2. e-07	2. e-09
0.1	x_4	3. e-06	2. e-07	x_5	8. e-09	1. e-09
0.1	x_5	3. e-05	2. e-07	x_1	2. e-07	2. e-08
0.1	x_7	5. e-04	2. e-05	x_9	1. e-05	4. e-07
0.5	x_1	1. e-03	2. e-04	x_1	8. e-05	9. e-06
0.5	x_8	9. e-06	4. e-07	x_5	1. e-06	7. e-08
0.5	x_4	1. e-06	3. e-07	x_5	3. e-09	2. e-09
0.5	x_5	1. e-05	3. e-07	x_7	2. e-06	6. e-08
0.5	x_7	2. e-03	1. e-04	x_9	1. e-04	6. e-06
1.0	x_1	5. e-03	6. e-04	x_1	3. e-04	4. e-05
1.0	x_3	8. e-05	3. e-06	x_9	1. e-05	1. e-06
1.0	x_4	5. e-06	2. e-07	x_5	2. e-09	1. e-09
1.0	x_5	1. e-04	3. e-06	x_1	2. e-05	9. e-07
1.0	x_7	5. e-03	5. e-04	x_9	3. e-04	3. e-05

7. ERROR REPRESENTATION

We are interested in finding the error $R'(x)$ in the N-th order differential quadrature

$$u'(x_i) = \sum_{j=1}^{N} a_{ij} u(x_j) + R'(x_i). \tag{1}$$

The well-known interpolation error $R(x)$ vanishes at N points x_j, j=1, N and has N continuous derivatives (provided that $u^{(N)}(x)$ is continuous). But it is not generally true that $R^{(N)}(x)$ is the N-th derivative of the above function. In order to obtain a practical error estimate we observe that by virtue of Rolle's theorem, $R'(x)$ has at least one distinct root $\bar{x}_j$ (j=1, N-1) in the open interval (x_j, x_{j+1}). We may also apply Rolle's theorem to the function

$$r(X) = R'(X) - b \prod_{j=1}^{N-1} (X - \bar{x}_j), \tag{2}$$

where b(x) is picked in such a way that r(x) = 0 for any fixed x distinct from the $\bar{x}_j$. So that $r^{(n-1)}(X)$ is continuous and has a root $\bar{x}$ in the interval containing x and the $\bar{x}_j$. From $r^{(n-1)}(\bar{x}) = 0$ follows that

$$b = u^{(N)}(\bar{x})/(N-1)! \tag{3}$$

Now substituting this value of b in r(x) = 0 we obtain the final result

$$R'(x) = \frac{u^{(N)}(\bar{x})}{(N-1)!} \prod_{j=1}^{N-1} (x-\bar{x}_j) \tag{4}$$

which is valid for all x.

The same sort of argument shows that the error R"(x) in the approximation $u_{xx} = A^2 u$ of Section 6 can be written as in Equation (5) with $x_j < \bar{x}_j < x_{j+2}$

$$R''(x) = \frac{u^{(N)}(\bar{x})}{(N-2)!} \prod_{j=1}^{N-2} (x-\bar{x}_j). \tag{5}$$

Assuming that all the x_j and x lie in an interval h and that in this interval $|u^{(N)}(x)| \leq K$, the bound on the error of Equations (4) and (5) is

$$R'(x)| \leq K \frac{h^{N-1}}{(N-1)!}, \tag{6}$$

$$R''(x)| \leq K \frac{h^{N-2}}{(N-2)!}. \tag{7}$$

8. HODGKIN-HUXLEY EQUATION

In 1963 A.L. Hodgkin and A.F. Huxley received a Nobel Prize (shared with J.C. Eccles) in physiology and medicine for their work in formulating a mathematical model of the squid giant axon. Their model describes excitation and propagation of the nerve impulse by means of a system of nonlinear differential equations. This set of simultaneous equations consist of three first-order ordinary differential equations and one partial differential equation of a parabolic type.

Hodgkin and Huxley considered the distribution of membrane current in a long axon to be the same as the distribution of electric current in a one dimensional cable. They thus arrived at the transmission equation for the cable, i.e., a partial differential equation of the form

$$\frac{1}{r}\frac{\partial^2 V}{\partial x^2} = I_{tot} = C\frac{\partial V}{\partial t} + I \tag{1}$$

where $V(x, t)$ and I_{tot} are membrane potential and total current, respectively and r and C are constants. The above equation represents the conservation of current at any point x along the axon, i.e., the total membrane current (I_{tot}) is the sum of the displacement current of membrane capacitance in the ionic current (I). The ionic current is mainly due to the movement of potassium and sodium ions (plus a leakage current) through the membrane. The functional dependence of ionic conductances (sodium, potassium and leakage) was described empirically by a system of three first order rate equations.

Numerical solutions of the above mathematical model was obtained by several authors. Since in 1952 it was not considered practical to solve the partial differential equation as it stands, Hodgkin and Huxley used two different kinds of constraints to reduce the equation (1) to an ordinary differential equation. In one case, (the "space-clamp" constraint), by means of special experimental procedures they kept the membrane potential the same along a finite length of axon i.e. $V = V(t)$ only. In the second case a uniform wave form was considered, i.e. $V(x, t) = v(x-\theta t)$ where θ is the conduction velocity. Here equation (1) is reduced to a second order ordinary differential equation with value of θ found by means of a trial and error procedure. In both approximations the numerical results agreed well with the corresponding experimental measurements. However, the only way to study the initiation of propagated impulses in an unconstrained axon is to solve the original partial differential equation.

In 1966 Cooley and Dodge treated the unconstrained axon, using a method of finite differences. In this method the length of axon is divided into numerous segments, δx in length, sufficiently short that membrane potential V can be considered uniform over a segment. It turned out that in order to get good agreement with experimental observation, δx must be chosen at least as small as 0.05 cm.

In this chapter, we employ the new method of differential quadrature to solve the nonlinear partial differential equation of the original Hodgkin-Huxley model for the squid axon. As we will see later, this new technique has several advantages over the conventional finite difference method. For one thing, δx can be taken as large as 0.25 cm and still provide good agreement with experimental data.

9. EQUATIONS OF THE MATHEMATICAL MODEL

The complete mathematical model of a squid axon at time t (in msec) and distance x along axon (in cm), is summarized by the following set of equations. Variables n, m and h are Hodgkin-Huxley conductance variables.

$$C \frac{\partial V}{\partial t} = \frac{1}{r} \frac{\partial^2 V}{\partial x^2} - I = I_{tot} - I \tag{1}$$

$$I = \bar{g}_{Na} m^3 h(V-V_{Na}) + \bar{g}_K n^4 (V-V_K) + g_L (V-V_L) = I(V) \tag{2}$$

$$\frac{d}{dt} n = \alpha_n (1-n) - \beta_n n \tag{3}$$

$$\frac{d}{dt} m = \alpha_m (1-m) - \beta_m m \tag{4}$$

$$\frac{d}{dt} h = \alpha_h (1-h) - \beta_h h \tag{5}$$

The definition and standard values of the constants for squid axon is given by:

a = radius of the axon (0.0238 cm).

R = specific resistance of axoplasm (34.5 ohm cm).

r = 2R/a.

C = specific membrane capacitance (1 μf/cm^2).

$\bar{g}_{Na}$ = max. sodium conductance (120 m mho/cm^2).

V_{Na} = sodium equilibrium potential (115 mV).

$\bar{g}_K$ = max. potassium conductance (36 m mho/cm^2).

V_K = potassium equilibrium Potential (-12 mV).

g_L = nonspecific leakage conductance (0.3 mmho/cm^2).

V_L = equilibrium potential of leakage current (10.6 mV).

The empirical rate constants α's and β's are evaluated for each according to the following functions:

$$\alpha_n = 0.01\ (V+10)/[\exp(\frac{V+10}{10}) -1]\ \varphi \tag{6}$$

$$\beta_n = 0.125\ \exp\ (\frac{V}{80})\ \varphi \tag{7}$$

$$\sigma_m = 0.1\ (V+25)/[\exp\ (\frac{V+25}{10}) -1]\ \varphi \tag{8}$$

$$\beta_m = 4\ \exp\ (\frac{V}{18})\ \varphi \tag{9}$$

$$\alpha_h = 0.07\ \exp\ (\frac{V}{20})\ \varphi \tag{10}$$

$$\beta_h = 1/[\exp\ (\frac{V+30}{10}) + 1]\ \varphi \tag{11}$$

$$\varphi = 3^{(T-6.3)/1C} \tag{12}$$

where T is the temperature of the fibre in degrees centigrade.

For all calculations the initial condition at t=0 corresponds to a uniform resting state, i.e.

$$V(x, t=0) = 0 \tag{13}$$

$$n(t=0) = \alpha_n(0)/(\alpha_n(0)+\beta_n(0)) \tag{14}$$

$$m(t=0) = \alpha_m(0)/(\alpha_m(0)+\beta_m(0)) \tag{15}$$

$$h(t=0) = \alpha_h(0)/(\alpha_h(0)+\beta_h(9)) \tag{16}$$

To study propagation and excitation of impulse, an axon is stimulated with an intracellular microelectrode at its midpoint, x=0. The length of axon is taken to be 2d and there is a symmetry with respect to the midpoint. If the amplitude of the stimulating current at x=0 is I_s, the total current at that point is

$$I_{tot}\,(x=0) = I_s + 2\frac{1}{r}\frac{\partial^2 V}{\partial x^2}\Big]_{x=0} \tag{17}$$

The second term in the above equation is the difference between the incoming and outgoing longitudinal current at midpoint with a factor of two for the symmetry of the axon. The last boundary condition is for the end point of the axon x=d. Since no current flows out of the end, we take

$$\frac{1}{r}\frac{\partial^2 V}{\partial x^2}\Big]_{x=d} = 0 \tag{18}$$

The system of equations (1 through 12) with the proper boundary conditions (13 through 18) must be solved simultaneously to find a solution to the membrane potential V at any point x and any time t > 0.

10. NUMERICAL METHOD

To apply the method of differential quadrature to the partial differential equation of an unconstrained axon (3.1), we consider the linear operator L of section 2 to be a second derivative operator. Then by analogy to equation (2.3) for any time t we have

$$\frac{\partial^2 V}{\partial x^2}\Big]_{x=x_i} \cong \sum_{j=1}^{N} a_{ij} V(x_j), \quad i=1, \ldots, N \tag{1}$$

The coefficients a_{ij} are determined using the unit cubic cardinal spline which satisfies the proper boundary conditions.

Using the above approximation, solution of the Hodgkin-Huxley partial differential equation reduces a set of ordinary differential equation of the form

$$C\frac{dV_i}{dt} = \frac{1}{r}\sum_{j=1}^{N} a_{ij}V_j - I(V_i) \tag{2}$$

$$I(V_i) = \bar{g}_{Na} m_i^3 h_i (V_i - V_{Na}) + \bar{g}_K n_i^4 (V_i - V_K) + g_L (V_i - V_L) \tag{3}$$

$$\frac{d}{dt} n_i = \alpha_n(1-n_i) - \beta_n n_i \quad (4)$$

$$\frac{d}{dt} m_i = \alpha_m(1-m_i) - \beta_m m_i \quad (5)$$

$$\frac{d}{dt} h_i = \alpha_h(1-h_i) - \beta_h h_i. \quad (6)$$

Where $V_i = V(x_i, t)$ and x_i is the distance along the axon from the stimulating electrode. Similarly, $n_i = n(V_i)$, $m_i = m(V_i)$ and $h_i = h(V_i)$.

One must note that the above approximation does not put any physical constraint upon the cable equation as in the case of space-clamp and uniform-wave models. Therefore, through the solution of the above set of equations, the initiation of propagated impulse in an unconstraint axon can be investigated.

Equations (2) through (6) form a set of four ordinary differential equation with boundary conditions given in Section 3, Equations (13) through (16). Numerical solution of this system of equations can be easily obtained using an Adams-Moulton integration scheme with a step size of $\delta t = 0.01$ msec.

11. CONCLUSION

The result of numerical calculation for a stimulus pulse of strength 11 μa and duration 0.2 msec at 18.5°C is plotted in Figure 1. The first plot shows the dependence of membrane potential V on the position x for various times t and the second plot gives V against t for various distances.

Comparison of differential quadrature technique with the uniform wave solution shows that:

(1) The two wave forms superimpose over most of the time course (within the accuracy of the plots). They agree well in peak amplitude (90.5mV) and duration. There is a slight discrepancy in rising phase of the solution.

(2) The predicted conduction velocity θ is well within the range of acceptable values:

θ for differential quadrature = 18.9 m/sec;
θ for uniform wave assumption = 18.8 m/sec;
θ for finite difference method = 18.67 m/sec.

Comparison with the finite difference method shows that this method is much simpler to employ and much faster to use on the digital computer. Step size along x-axis in this method could be as large as $\delta x=0.5$ cm and still produce good results, whereas in finite difference method max $(\delta x) = 0.05$ cm.

(3) The effect of the cable properties on the repetitive firing of the continuous axon was also examined. The partial differential equation was solved for step stimuli of various amplitudes. Solutions are in good agreement with the results of Cooley and Dodge.

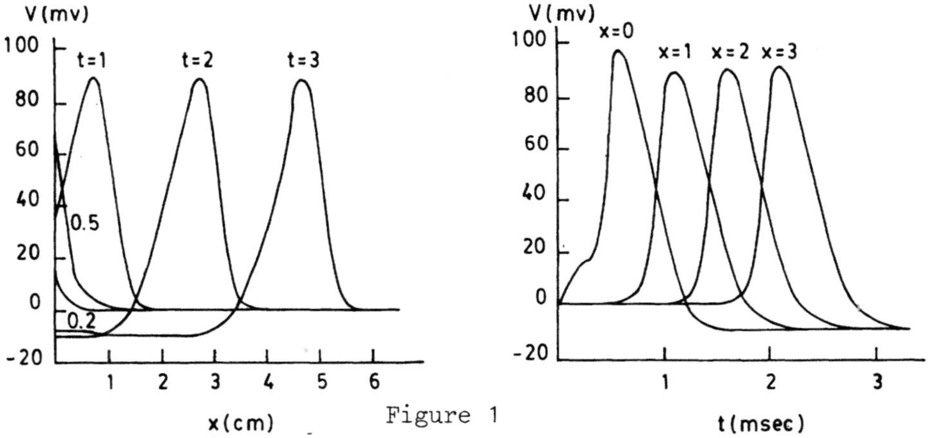

Figure 1

BIBLIOGRAPHY AND COMMENTS

Section 1. We are following the paper:

R. Bellman, B.G. Kashef, and J. Casti, 'Differential Quadrature: A Technique for the Rapid Solution of Non-linear Partial Differential Equations', Journal of Computational Physics 10 (1972) 40-52.

Section 8. These results were given in:

R. Bellman and B. Kashef, 'Solution of the Partial Differential Equation of the Hodgkin-Huxley Model using Differential Quadrature', Mathematical Biosciences 19 (1974) 1-8.

For further results and many additional references, see:

S. Kaplan and D. Trujillo, 'Numerical Studies of the Partial Differential Equations Governing Nerve Impulse Conduction', Mathematical Biosciences 7 (1970).

Chapter XIII

ADAPTIVE GRIDS AND NONLINEAR EQUATIONS

1. INTRODUCTION

Since the equations of hydrodynamics are non-linear, a computational solution of these equations usually involves a numerical integration. This numerical integration is particularly difficult to perform in the presence of discontinuities of the solution called "shocks". Although there are a number of devices available for obtaining the solution in the neighborhood of the shock, notably by von Neumann, Richtmeyer and Lax, these have the unfortunate property of altering the behavior of the solution in other regions. If conventional finite difference schemes are employed, instabilities arise in a number of cases.

In this chapter we propose a new method of numerical integration which seems well suited to non-linear hyperbolic equations of the form

$$u_t = g(u)u_x + h(u, x, t), \qquad u(x, 0) = f(x). \tag{1}$$

To illustrate the method, we consider the pseudo-hydrodynamic equation

$$u_t = -uu_x, \qquad u(x, 0) = f(x), \tag{2}$$

which exhibits the shock phenomenon and which possesses the great merit of an analytic solution. We can thus compare the analytic solution with that obtained from the application of our numerical technique.

2. THE EQUATION $u_t = -uu_x$

The equation appearing in (1.2) has the solution

$$u(x, t) = f[x-u(x, t)t]. \tag{1}$$

Since

$$u_x(x, t) = \frac{f'(E)}{1+tf'(E)}, \quad E=x-u(x, t)t \tag{2}$$

it is clear that u_x becomes infinite whenever $1+tf'(E) = 0$. This is the shock phenomenon. Numerical integration is difficult in the neighborhood of this shock.

In place of the usual finite difference method based upon a fixed (x, t) - grid, consider the following scheme:

$$u(x, t+\Delta) = u[x-u(x, t)\Delta, t], \quad u(x, 0) = f(x) \tag{3}$$

where t assumes the values $0, \Delta, 2\Delta, \ldots,$. Expansion of the two sides in powers of Δ reveals that (3) is equivalent to (1.2) to first order terms in Δ.

The method does not involve the strict notion of an (x, t) grid, since the $x-u(x, t)\Delta$ values will not be distributed in any regular fashion. In integrating the equation, a fixed x-grid is used,
$x = \ldots, -2\delta, -\delta, 0, \delta, 2\delta, \ldots$.
When the values $u(n\delta, t)$ are known for fixed t the quantities

$$u[n\delta - u(n\delta, t)\Delta, t] = u(n\delta, t+\Delta)$$

are found by interpolation in the x direction. There is no constraint on the relation between δ and Δ, and δ may be chosen as small as desired, subject to space limitations on the digital computer and the cost of computing time.

One important feature of the recurrence scheme in (3) above is that uniform boundedness of $u(x, t)$ over x for $t = k\Delta$ is automatically passed on to $u(x, t)$ at $t=(k+1)\Delta$.

3. AN EXAMPLE

We have integrated (2.3) using for the starting condition

$$f(x) = \begin{cases} 0, & -1.0 \le x \le -0.5; \\ 5/12(x+\frac{1}{2}), & -0.5 \le x \le -0.26; \\ 0.1, & -0.26 \le x \le 0.26; \\ 5/12(\frac{1}{2}-x), & 0.26 \le x \le 0.5; \\ 0 & 0.5 \le x \le 1.0; \end{cases} \tag{1}$$

$$f(x) = f(x+2), \quad -\infty < x < \infty.$$

We have used six-point Lagrangian interpolation, x points at intervals of $\delta = 0.02$, and various values of Δ. Some results are shown in Table I. It should be noted that even with relatively large Δ the discontinuity in u that takes place at $t = 2.4$ is handled quite well. Other starting conditions have also proved successful.

4. DISCUSSION

Since the method consists of integrating along the characteristic, good results are to be expected. However, we have also integrated equations of the form

$$t_u = g(u)u_x + h(u, x, t) \qquad \qquad (1)$$
$$u(x, 0) = f(x), \qquad -\infty < x < \infty,$$

using the difference scheme

$$u(x, t+\Delta) = u(x-g[u(x, t)]\Delta, t)+h[u(x, t), x, t]\Delta, \qquad (2)$$
$$u(x, 0) = f(x), \quad -\infty < x < \infty,$$

where the characteristic curves are not simply $x-g(u)t = k$, and have obtained excellent results.

5. EXTENSION

The method is applicable to the integration of the system

$$u_t = G(u, v, x, y)+H(u, v, x, y)u_y+g(u, v, x, y), \qquad (1)$$
$$v_t = M(u, v, x, y)u_x+N(u, v, x, y)v_y+h(u, v, x, y).$$

The corresponding difference scheme is

$$u(x, y, t+\Delta) = u\{x+G\Delta, y+H\Delta, t\}+g\Delta \qquad (2)$$
$$u(x, y, t+\Delta) = v\{x+M\Delta, y+N\Delta, t\}+h\Delta.$$

It is readily verified that (5.2) is equivalent to (5.1) to terms of order Δ. The integration proceeds as in the previous case, a grid in the (x, y) plane being chosen and interpolation among these grid points being used to advance the integration in time.

However, attempts at extending the integration method in

an obvious way to general systems of hyperbolic equations in one space variable have not been successful. While the method presented here is marginally stable, the attempted generalizations have suffered from instabilities.

TABLE I
Values of u(x, t) at t=2.4

x	Exact u	Computed u			
		Δ=0.02	Δ=0.10	Δ=0.20	Δ=0.30
-0.48	0.004167	0.003713	0.003657	0.003584	0.003508
-0.40	0.0208	0.020805	0.02056	0.02024	0.01991
-0.32	0.0375	0.03738	0.03694	0.03637	0.03570
-0.24	0.0542	0.05390	0.05336	0.05253	0.05165
-0.16	0.0708	0.07063	0.06981	0.06809	0.06745
-0.08	0.0875	0.08748	0.08621	0.08481	0.08255
0.00	0.1000	0.10022	0.10034	0.10000	0.10072
+0.08	0.1000	0.09994	0.10000	0.10000	0.10002
-0.16	0.1000	0.10000	0.10000	0.10000	0.10000
-0.24	0.1000	0.10000	0.10000	0.10000	0.10000
-0.32	0.1000	0.10002	0.10000	0.10000	0.10000
-0.40	0.1000	0.10028	0.10008	0.10000	0.10009
-0.44	0.1000	0.09924	0.09952	0.10000	0.10111
-0.46	0.1000	0.09785	0.10351	0.09884	0.08966
-0.48	0.1000	0.12222	0.08834	0.06096	0.04563
+0.50	-	0.00000	0.00000	0.00000	0.00000
+0.52	0.0000	0.00000	0.00000	0.00000	0.00000
+0.54	0.0000	0.00000	0.00000	0.00000	0.00000
+0.56	0.0000	0.00000	0.00000	0.00000	0.00000

6. HIGHER ORDER APPROXIMATIONS

To obtain higher order approximations we may modify the difference scheme using the differential equation. This will involve the function and its first derivative, but the storage problems are minimal.

BIBLIOGRAPHY AND COMMENTS

Section 1. This method was given in:

R. Bellman, I. Cherry, and G.M. Wing, 'A Note on the Numerical Integration of a Nonlinear Hyperbolic Equation', Quarterly of Applied Mathematics 16 (1958) 181-183.

The use of adaptive grids for ordinary differential equations was discussed in

R. Bellman, Adaptive Control Processes: A Guided Tour, Princeton University Press, Princeton, New Jersey, 1961.

Chapter XIV

INFINITE SYSTEMS OF DIFFERENTIAL EQUATIONS

1. INTRODUCTION

One of the most tantalizing areas of modern mathematical physics is the theory of turbulence. The complexity of the physical process, combined with the nonlinearity of the hydrodynamical equations, precludes any direct analytic approach and forces us to study various types of approximation techniques. A quite useful technique in many investigations is the analysis of "model equations" which, hopefully, exhibit many of the essential characteristics of the more realistic and correspondingly more recalcitrant equations.

Of these model equations in turbulence, perhaps the best known is the equation of Burgers,

$$u_t + uu_x = \varepsilon u_{xx}, \quad x(x, 0) = g(x), \tag{1}$$

which, as shown by Hopf and others, can be reduced to the linear heat equation by a change of variable and thus can be solved explicitly. Since the analytic solution is available, this equation furnishes a very important test of any proposed analytic approximation method or of any computational procedure for the actual hydrodynamic equations.

In this chapter we wish to present the results of two computational methods which have been made feasible only in recent years with the development of the high speed digital computer. With the present greater speeds and larger rapid-access storage, we can expect that routine algorithms can be applied to the study of many important physical processes.

In the first part of this chapter, we consider an application of the classical method of converting a partial differential equation into an infinite system of ordinary differential equations. This method was studied by Lichtenstein, Siddiqu and Minakshisundaram.

Fifty years ago, the possibility of using this technique to obtain numerical solutions of partial differential equations was slight, since in order to obtain sufficient accuracy one has to think in terms of 50, 100 or 1000 equations. As we shall see below, the number of equations accuracy depends upon the value of ε in (1). With modern computers, the prospects are quite different.

There are several advantages in reducing a numerical process to the solution of ordinary differential equations with initial conditions. First, current computers are ideally designed for this type of repetitive process. Second, we now have available a large stock of computational algorithms of guaranteed accuracy for the solution of equations of this type. This makes the calculation routine, since we have transformed the original problem into that of solving a system of the form

$$\frac{dx_i}{dt} = g_i(x_2, x_2, \ldots, x_N), \quad x_i(0)=c_i, \quad i=1, 2, \ldots, N. \tag{2}$$

There is, however, much to be gained from effecting this transformation efficiently. This brings us to the subject of "closure", one of the basic problems of mathematical physics. In this chapter we further indicate how the techniques may be applied, and show how extrapolation and nonlinear summability methods are used in this case.

We also discuss the use of some new types of difference approximation which have already proved themselves in other connections.

Numerical results are given throughout, together with the time required for computing each case.

At the end of the chapter, we discuss some of the stability questions connected with truncation.

2. BURGERS' EQUATION

Let us consider (1.1) over the region $-\infty < x < \infty$, $t > 0$, and suppose, for the sake of simplicity, that $g(x)$ is periodic, with period 2π. We wish then to find a periodic solution of (1).

$$u(x, t) = \sum_{-\infty<k<\infty} u_k(t)e^{ikx} = \sum_{-\infty<k<\infty} u_k(t)(\cos kx+i \sin kx) \tag{1}$$

and $u_k(t) = v_k(t) + iw_k(t)$, where v_k and w_k are real. The condition that $u(x, g)$ be real requares that

$$v_k = v_{-k}, \quad w_k = -w_{-k}. \tag{2}$$

Substituting the expression in (1) into (1.1) and equating coefficients, we obtain the infinite system of ordinary differential equations

$$u_n'(t) + \sum_{-\infty<k<\infty} iku_k(t)u_{n-k}(t) = -n^2\varepsilon u_n(t). \tag{3}$$

Here $\varepsilon > 0$. Substituting $u_n = v_n + iw_n$ and equating real and imaginary parts, we obtain the real system

$$\begin{aligned} v_n' - \sum_{-\infty<k<\infty} k(w_k v_{n-k} + v_k w_{n-k}) &= -n^2\varepsilon v_n, \\ w_n' + \sum_{-\infty<k<\infty} k(v_k v_{n-k} - w_k w_{n-k}) &= -n^2\varepsilon w_n. \end{aligned} \tag{4}$$

Since this is an infinite system of equations, we must use some closure technique in order to obtain a finite system. At this point we employ the simple device of setting

$$v_k, w_k = 0, \quad |k| \geq N+1 \tag{5}$$

for some value of N. Taking various values of N and ε, we explore the degree of accuracy obtained as a function of these variables, by carrying out a computational solution of the resulting finite system of ordinary differential equations.

The finite system of differential equations obtained from (4), using the foregoing closure technique, is

$$\begin{aligned} v_n' - \sum_{k=0}^{n} k(w_k v_{n-k} + z_k w_{n-k}) - \sum_{k=n+1}^{N} k(w_k v_{k-n} - v_k w_{k-n} + \\ + \sum_{k=1}^{N-n} k(-w_k v_{n+k} + v_k w_{n+k}) = -n^2\varepsilon v_n, \\ w_n' + \sum_{k=0}^{n} k(v_k v_{n-k} - w_k w_{n-k}) + \sum_{k=n+1} k(v_k v_{k-n} + w_k w_{k-n}) - \\ - \sum_{k=1}^{N-n} k(v_k v_{n+k} + w_k w_{n+k}) = -n^2\varepsilon w_n, \end{aligned} \tag{6}$$

for n=0, 1, 2, ..., N. This is a system of 4N simultaneous equations with initial values.

$$g(x) \sim \sum_{-\infty<k<\infty} g_k e^{ikx}. \tag{7}$$

That is if $g_k = h_k + ir_k$, we have $v_k(0) = h_k$, $w_k(0) = r_k$.

3. SOME NUMERICAL EXAMPLES

A FORTRAN program was written for the IBM 7090 in order to study the effect of varying N and ε in (2.6). As mentioned above, the minimum value of N required to give accurate values of, for example, $u_1(t)$, should depend on ε. The integration was carried out using either the fixed step Runge-Kutta or the variable step Adams-Moulton technique.

Three cases were considered

Case I:

$$\begin{aligned} &(a) \quad \varepsilon = 0.01, \\ &(b) \quad u_1(0) = 1,\ u_k(0) = 0,\ k \neq 1, \\ &(c) \quad N = 5,\ 10,\ 25,\ 36,\ 50. \end{aligned} \tag{1}$$

Figure 1 shows $v_1(t)$ as a function of t for different values of N. A cutoff at N = 36 yields results substantially as good as those for N = 50, in half the computation time.

Case II:

$$\begin{aligned} &(a) \quad \varepsilon = 0.1, \\ &(b) \quad u_1(0) = 1,\ u_k(0) = 0,\ k \neq 1, \\ &(c) \quad N = 5,\ 10,\ 25. \end{aligned} \tag{2}$$

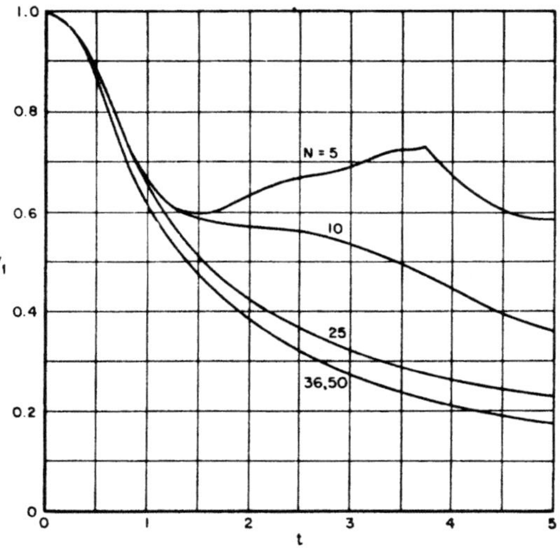

Figure 1
$v_1(t)$ as a function of t for N=5, 10, 25, 36, 50,
for $\varepsilon = 0.01$

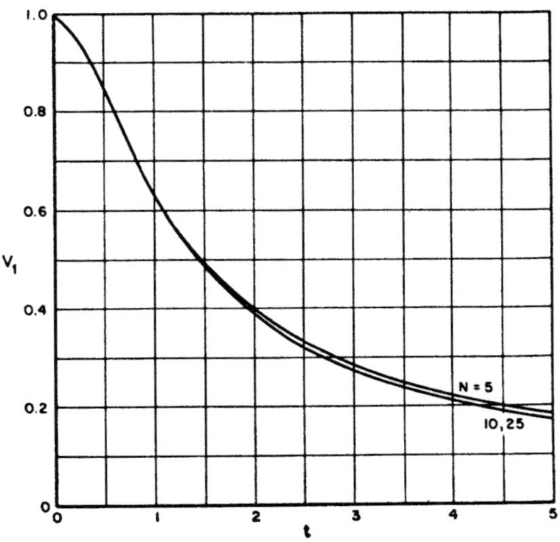

Figure 2
$v_1(t)$ as a function of t for N=10, 25
for $\varepsilon = 0.1$

This example shows the effect of increasing the value of ε, since significantly accurate results are obtained using a cutoff of N=6. N=10 and 25 give the same curve (see Figure 2).

Case III:

$$\begin{aligned} &(a) \quad \varepsilon = 0.01, \\ &(b) \quad u_1(0) = i, \; u_k(0) = 0 \quad \text{for } k \neq 1, \\ &(c) \quad N = 5, 10, 25, 36, 50. \end{aligned} \tag{3}$$

Computed values for w_i can be compared in Table I.

TABLE I

w_1	N=5	N=10	N=25	N=36	N=50
t=1.1	0.634	0.640	0.618	0.612	0.612
t=2.2	0.654	0.568	0.398	0.375	0.370
t=3.3	0.718	0.513	0.306	0.269	0.264
t=4.4	0.614	0.407	0..255	0.210	0.205
t=4.95	0.591	0.367	0.236	0.189	0.184
computer time	1/6 min	1 min	2 min	5 min	12 min

The calculation N=36 is accurate to ± 0.005 compared with N=50.

Figures 3, 4, and 5 show the dependence of $w_k(t)$ on t = 1, 2, 3, 4, 5, as k varies, where N=36.

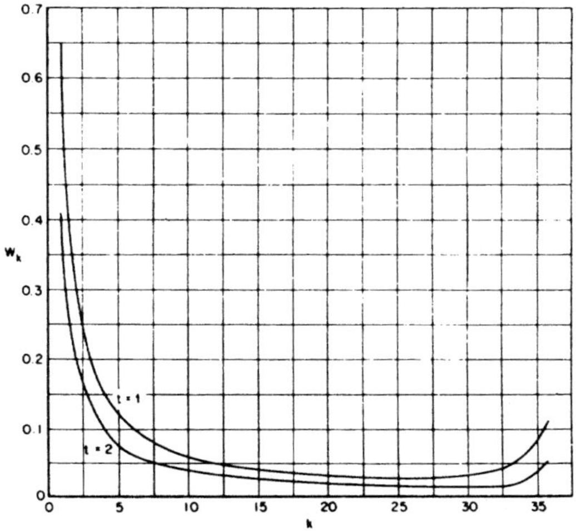

Figure 3
The dependence of $w_k(t)$ on k for t=1, 2, with N=36, ε=0.01

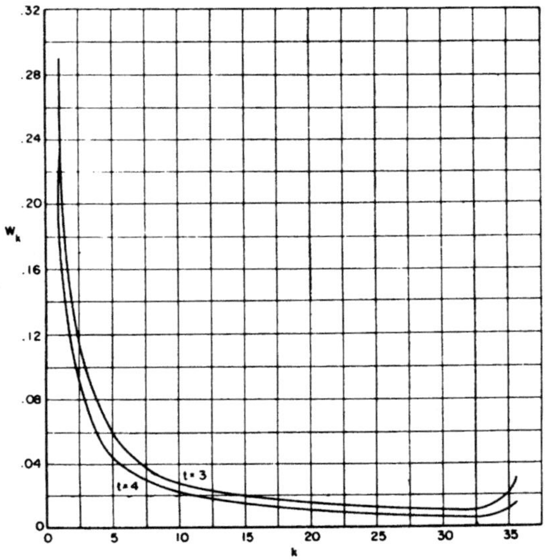

Figure 4
The dependence of $w_k(t)$ on k for t=3, 4, with N=36, ε=0.01

4. TWO-DIMENSIONAL CASE

For the two-dimensional case, we wish to solve the system of partial differential equations,

$$\begin{aligned} u_t + uu_x + vu_y &= (u_{xx} + u_{yy}), \\ v_t + uv_x + vv_y &= (v_{xx} + v_{yy}), \end{aligned} \tag{1}$$

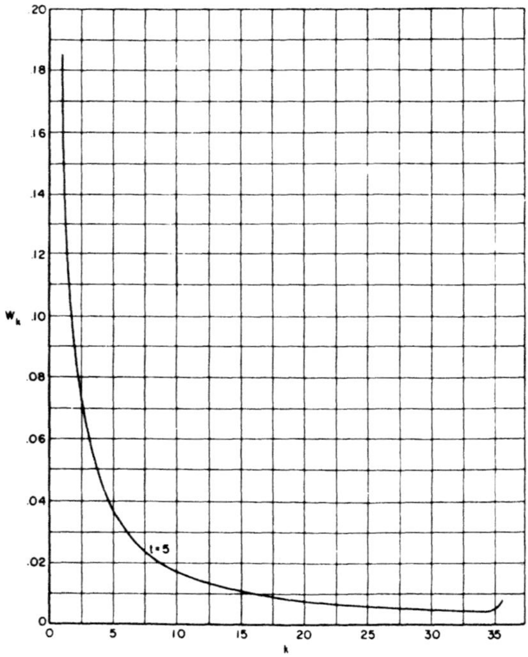

Figure 5
The dependence of $w_k(t)$ on k for t=5, with N=36, ε=0.01

obvious analogs of the equation of Burgers. Unfortunately, these equations cannot be solved in closed analytic form for ε > 0. For ε = 0, the equations can be solved. Once again, we represent the solutions as a Fourier series

$$
\begin{aligned}
u(x, y, t) &= \sum_k \sum_\ell u_{k\ell}(t)e^{i(kx+\ell y)}, \\
v(x, y, t) &= \sum_k \sum_\ell v_{k\ell}(t)e^{i(kx+\ell y)},
\end{aligned} \tag{2}
$$

where $u_{k\ell} = a_{k\ell} + ib_{k\ell}$, $v_{k\ell} = c_{k\ell} + id_{k\ell}$.

Substituting (1), we obtain the following infinite systems of differential equations:

$$
u'_{m,n}(t) + \sum_k \sum_\ell iku_{k,\ell}u_{m-k,n-\ell} + \sum_k \sum_\ell i\ell u_{k,\ell}v_{m-k,n-\ell} = -\varepsilon(m^2 + n^2)u_{m,n}, \tag{3}
$$

$$
v'_{m,n}(t) + \sum_k \sum_\ell ikv_{k,\ell}u_{m-k,n-\ell} + \sum_k \sum_\ell i\ell v_{k,\ell}v_{m-k,n-\ell} = -\varepsilon(m^2 + n^2)v_{m,n}, \tag{4}
$$

where $-\infty \leq k, \ \ell < +\infty$.

Substituting $u_{k\ell} = ak_\ell$ and $v_{k\ell} = c_{k\ell} + id_{k\ell}$ in (4), and separating into real and imaginary parts, we obtain an infinite set of real ordinary differential equations. If we assume that $u_{mn} = v_{mn} = 0$ for $|m|, |n| \geq N+1$ for some integer $N > 0$, we again have a finite system:

$$
\begin{aligned}
a'_{m,n} &- \sum_k \sum_\ell k(a_{\alpha,\beta}b_{k,\ell} + a_{k,\ell}b_{\alpha,\beta}) - \\
&- \sum_k \sum_\ell \ell(b_{k,\ell}c_{\alpha,\beta} + a_{k,\ell}d_{\alpha,\beta}) = -\varepsilon a_{m,n}(m^2+n^2), \\
b'_{m,n} &+ \sum_k \sum_\ell k(a_{k,\ell}a_{\alpha,\beta} - b_{k,\ell}b_{\alpha,\beta}) + \\
&+ \sum_k \sum_\ell \ell(a_{k,\ell}c_{\alpha,\beta} - b_{k,\ell}d_{\alpha,\beta}) = -\varepsilon b_{m,n}(m^2+n^2), \\
c'_{m,n} &- \sum_k \sum_\ell k(d_{k,\ell}a_{\alpha,\beta} + c_{k,\ell}b_{\alpha,\beta}) - \\
&- \sum_k \sum_\ell \ell(d_{k,\ell}c_{\alpha,\beta} + c_{k,\ell}d_{\alpha,\beta}) = -\varepsilon c_{m,n}(m^2+n^2),
\end{aligned}
$$

$$d'_{m,n} + \sum_k \sum_\ell k(c_{k,\ell} a_{\alpha,\beta} + d_{k,\ell} b_{\alpha,\beta}) +$$

$$+ \sum_k \sum_\ell \ell(c_{k,\Omega} c_{\alpha,\beta} - d_{k,\ell} d_{\alpha,\beta}) = -\varepsilon d_{m,n}(m^2+n^2),$$

where $\alpha = m-k$, $\beta=n-\ell$, $-N \leq k$, and $\ell < N$.

We can now solve this system for positive and negative indices, in which there are $(2N+1)^2$ coefficients, or, we can reduce the number of coefficients by utilizing the symmetry relations:

$$a_{-m,-n} = a_{m,n} \qquad b_{-m,-n} = -b_{m,n}$$

$$a_{-m,n} = a_{m,-n} \qquad b_{-m,n} = -b_{m,-n}$$

$$c_{-m,-n} = c_{m,n} \qquad d_{-m,-n} = -d_{m,n}$$

$$c_{-m,n} = c_{m,-n} \qquad d_{-m,n} = -d_{m,-n}$$

If we follow the latter procedure, each coefficient is written in terms of 12 double sums. For programming purposes, the first method is preferable.

We have carried out some preliminary calculations testing the feasibility of the method. Unfortunately, the times involved make it difficult to carry out the type of experimentation discussed in the foregoing sections in connection with the one-dimensional equation. The situation has greatly improved in the past year; new computers will cut down the required time by at least a factor of 10.

5. CLOSURE TECHNIQUES

The general question of closure may be phrased in the terms given below. Starting with a system of differential equations,

$$\frac{dx}{dt} = g(x) \tag{1}$$

in vector notation, we want to replace it by another system

$$\frac{dy}{dt} = h(y), \tag{2}$$

with simpler analytic, conceptual, and computational properties, so that $\|x-y\|$ is small. Here $\|\ldots\|$ is some suitably defined norm; for analytic convenience it is usually the Euclidean norm. In some cases, we are content to have $h(y)$ linear, i.e., $h(y) = Ay + b$ but in other cases we are more interested in having the dimension of y considerably less than the dimension of x.

Here we wish to consider an infinite-dimensional case where (1) corresponds to the system of (2.4). The problem of closure is now more difficult since there is no immediate way of expressing the components corresponding to higher harmonics in terms of the components for smaller k.

We discuss two approaches in Sections 6 and 7 below.

6. A DIRECT METHOD

Consider the equation

$$u_t + uu_x = \varepsilon u_{xx}, \tag{1}$$

where we set

$$u(x, t) = \sum_{-\infty<k<\infty} u_k(t)e^{ikx}. \tag{2}$$

To obtain a finite set of differential equations, set

$$u = \sum_{|k|\leq N} u_k(t)e^{ikx}, \tag{3}$$

substitute in (1) and equate coefficients.

7. EXTRAPOLATION

We now present another technique. Suppose that we have already employed Carleman linearization to obtain an infinite linear system of the form

$$\frac{du_k}{dt} = \sum_{\ell=1}^{\infty} a_{k\ell}u_\ell, \quad u_k(0) = c_k, \quad k=1, 2, \ldots, \tag{1}$$

and suppose further that we wish to obtain an approximating linear system of the form

$$\frac{du_k}{dt} = \sum_{\ell=1}^{N} \alpha_{k\ell} u_\ell, \qquad u_k(0) = c_k. \tag{2}$$

Instead of merely choosing $\alpha_{k\ell} = \alpha_{k\ell}$, we wish to approximate to the remainder terms by means of linear combinations of the initial terms

$$\sum_{\ell \geq N+1} a_{k\ell} u_\ell \simeq \sum_{\ell=1}^{N} b_{k\ell} u_\ell. \tag{3}$$

The coefficients $b_{k\ell}$ are to be chosen so that

$$\int_0^T \left[\sum_{\ell \geq N+1} a_{k\ell} u_\ell - \sum_{\ell=1}^{N} b_{k\ell} \right]^2 dt \tag{4'}$$

is a minimum. In this way we obtain linear algebraic equations for the $b_{k\ell}$:

$$\sum_{\ell \geq N+1} a_{k\ell} \int_0^T u_\ell u_r \, dt = \sum_{\ell=1}^{N} b_{r\ell} \int_0^T u_\ell u_r \, dt, \qquad r=1, 2, \ldots, N. \tag{5}$$

The usual difficulty now confronts us. How do we compute the integrals $\int_0^T u_\ell u_r \, dt$ involving the unknown solution?

Consider first the case where $1 < r, \ell \leq N$. To obtain these integrals, we use the finite system in (2), with $\alpha_{k\ell} = a_{k\ell}$. Call the solutions $u_r^{(0)}$. The coefficients of $b_{r\ell}$ in (5) are then $\int_0^T u_\ell^{(0)} u_r^{(0)} \, dt$. Observe that these quantities can be computed directly in the course of obtaining the $u_k^{(0)}$ by adjoining to (2) the equations

$$\frac{dw_{r\ell}}{dt} = u_r u_\ell, \qquad w_{r\ell}^{(0)} = 0, \tag{6}$$

and seeking only the values $w_{r\ell}(T)$.

The more difficult problem is that of the calculation of $\int_0^T u_r u_\ell \, dt$ for r=1, 2, ..., w, $\ell \geq N+1$. Here we use extrapolation techniques. Keep r fixed and let ℓ vary over the integers, ℓ=1, 2, ... It is reasonable to expect that $f_{\ell, r} = \int_0^T u_r^{(0)} u_\ell^{(0)} \, dt$ will be a well-behaved sequence. Consequently, if we possess the values $f_{1, r}, f_{2, r}, \ldots, f_{N, r}$, we can use any of a number of extrapolation techniques to obtain the values of $f_{\ell, r}$ for $\ell \geq N+1$.

The system in (5) then has the form

$$\sum_{M \geq \ell \geq N+1} a_{k\ell} f_{\ell, r} = \sum_{\ell=1}^{N} b_{r\ell} w_{r\ell}(T). \tag{7}$$

Here M is a cutoff number, such as 2N, which depends upon the size of the coefficients $a_{k\ell}$ and the rapidity of convergence of the infinite series.

Solving (7) numerically, we obtain the coefficients $b_{r\ell}^{(0)}$. We use the superscript to indicate the fact that these are the first approximations. To obtain higher approximations, we use self-consistency techniques.

In place of (2), with $\alpha_{k\ell} = a_{k\ell}$ let us now use the system

$$\frac{du_k}{dt} = \sum_{\ell=1}^{N} a_{k\ell} u_\ell + \sum_{\ell=1}^{N} b_{k\ell}^{(0)} u_\ell, \qquad u_k^{(0)} = c_k. \tag{8}$$

Call the solutions of this equation $u_k^{(1)}$, k=1, 2, ..., N. We now proceed as before to calculate $w_{r, \ell}^{(1)}(T)$, $f_{\ell, r}^{(1)}$ and $b_{k\ell}^{(1)}$. With the new coefficients $b_{k\ell}^{(1)}$, we introduce the equation

$$\frac{du_k}{dt} = \sum_{\ell=1}^{N} a_{k\ell} u_\ell + \sum_{\ell=1}^{N} b_{k\ell}^{(1)} u_\ell, \; u_k^{(0)} = c_k. \tag{9}$$

This process is repeated for a fixed number of steps or until the values of the $b_{k\ell}$ settle down.

8. DIFFERENCE APPROXIMATIONS

An entirely different approach to the numerical solution of partial differential equations is based upon the use of difference euqations. In an earlier chapter we introduced a new type of difference equation technique. Here we wish to apply it to the Burgers' equation.

9. AN APPROXIMATING ALGORITHM

Consider the equation (1.1) over the region $0 \leq x \leq 1$, $t > 0$, and suppose that $g(x)$ is periodic, with period π. Let the approximating algorithm be given by

$$u(x,\ t+\Delta) = \lambda u(x-au(x,\ t)\Delta,\ t)+ \frac{(1-\lambda)}{2}[u(x+b\Delta^{\frac{1}{2}},\ t) + u(x-b\Delta^{\frac{1}{2}},\ t)], \tag{1}$$

where Δ is the integration step size, and λ, a, b are constants which will be determined. To show that (1) approximates (1.1) to an error of $O(\Delta^2)$, expand both sides of (1) in a Taylor series up to the Δ^2 term, obtaining the equation

$$u_t = -\lambda a u u_x + (1-\lambda)\frac{b^2}{2} u_{xx}. \tag{2}$$

For (2) to approximate (1.1), the following relations must hold:

$$a = 1/\lambda, \qquad b = \left(\frac{2\varepsilon}{1-\lambda}\right)^{\frac{1}{2}}. \tag{3}$$

If ε is fixed, then a and b are functions of the parameter λ, and (1) becomes

$$u(x,\ t+\Delta) = \frac{1}{\lambda}\, u\left[x-u(x,\ t)\frac{\Delta}{\lambda},\ t\right] + \frac{1-\lambda}{2}\left[u\left(x+\left(\frac{2\varepsilon\Delta}{1-\lambda}\right)^{\frac{1}{2}},\ t\right) + u\left(x-\left(\frac{2\varepsilon\Delta}{1-\lambda}\right)^{\frac{1}{2}},\ t\right)\right]. \tag{4}$$

Let $t=0,\ \Delta,\ 2\Delta,\ \ldots,$ and at each stage of the calculation let $u(x,\ t)$ be stored by means of the finite sum

$$u(x,\ t) \cong \sum_{n=1}^{M} u_n(t) \sin n\pi x, \tag{5}$$

where the coefficients $u_n(t)$ are obtained by the quadrature scheme

$$u_n(t) = 2\int_0^1 u(x, t) \sin n\pi x \, dx \simeq \frac{2}{R} \sum_{k=1}^{R-1} u(k/R, t) \sin (n\pi k/R). \tag{6}$$

Hence, the values $u(k/R, t)$, $k=1, 2, \ldots, R-1$ store $u(x, t)$ at time t, and by way of (4), $u(x, t+\Delta)$ can be obtained.

10. NUMERICAL RESULTS

To obtain some numerical results, a FORTRAN program was written for the IBM 7090. The following results were obtained:

$$\text{(A)}\quad u_t + uu_x = \varepsilon u_{xx}, \tag{1}$$

where

$$u(x, 0) = -\sin \pi x, \quad 0 \le x \le 1$$

$$\varepsilon = 0.01, \quad \lambda = 0.5, \quad \Delta = 0.05.$$

In this first example the parameters M and R were varied. As can be seen in Table II, one obtains essentially the same results for M = R = 10, compared with M = R = 15, in less than half the time.

TABLE II

x	t	u(x, t)	
		M = R = 10	M = R = 15
0.5	1.00	-0.355	-0.354
0.1	2.00	-0.319	-0.326
0.6	3.00	-0.116	-0.116
0.9	4.00	-0.022	-0.022
0.2	5.00	-0.132	-0.133
Time		40 sec	1 min

Note that in this example and the following examples the largest differences occur for small values of x.

(B) $u_t + uu_x = \varepsilon u_{xx}$,

where

$$u(x, 0) = -\sin \pi x, \quad \varepsilon \leq x \leq 1$$

$$\varepsilon = 0.01, \quad \Delta = 0.95, \quad m = R = 10.$$

In the second example the parameter λ was varied. As shown in Figure 6, varying λ displays large differences in the values of $u(x, t)$ for small x and for small t. Note, however, that the maximum of each curve occurs at about the same time. For small t and large x, the variation in $u(x, t)$ is much smaller (see Figure 7); for large values of te, the variation is quite small (see Table III).

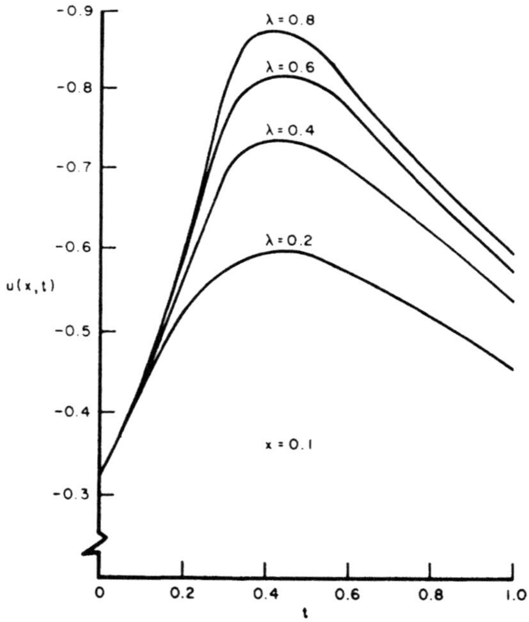

Figure 6
$u(x, t)$ as a function of t and λ, with x = 0.1

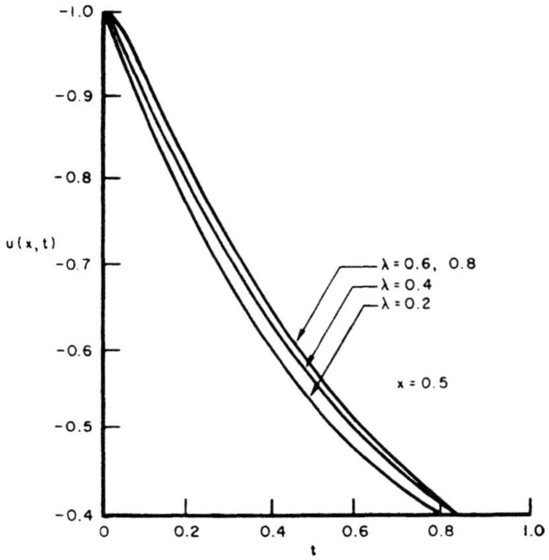

Figure 7
u(x, t) as a function of t and λ, with x = 0.5

TABLE III

x	t	λ = 2	λ = 8	difference
0.1	3	-0.184	-0.233	0.039
0.6	3	-0.115	-0.115	0.000
0.1	4	-0.313	-0.157	0.026
0.6	4	-0.089	-0.089	0.000
0.1	5	-0.098	-0.115	0.017
0.6	5	-0.73	-0.073	0.000

(C) $u_t + uu_x = \varepsilon u_{xx}$,

where

$$u(x, 0) = -\sin \pi x, \quad 0 \leq x \leq 1$$

$$\varepsilon = 0.01, \quad \lambda = 0.5, \quad M = R = 10.$$

The parameter which has the greatest effect on the results in the integration step size Δ. The difference can be seen in Table IV.

TABLE IV

x	t	u(x, t) $\Delta = 0.05$	u(x, t) $\Delta = 0.01$
0.5	1.0	-0.355	-0.381
0.1	2.0	-0.319	-0.375
0.6	3.0	-0.116	-0.120
0.9	4.0	-0.022	-0.023
0.2	4.8	-0.139	-0.147
Time		40 sec	180 sec

11. HIGHER ORDER APPROXIMATION

The general form for the approximating algorithm is given by

$$u(x, t+\Delta) = \lambda u(x-au(x, t)\Delta, t) + \sum_{i=1}^{N} a_i[u(x+b_i\Delta^{\frac{1}{2}}, t) + u(x-b_i\Delta^{\frac{1}{2}}, t)] \tag{1}$$

where λ, R, a_i, b_i, and a are inputs. It was shown that higher ordered approximations, as in (1), can give more accurate results, provided the polynomial approximation (9.5) is sufficiently accurate.

12. TRUNCATION

Consider the nonlinear partial differential equation

$$u_t = u_{xx} + u^2, \tag{1}$$

where u is subject to the initial condition

$$u(x, 0) = g(x),$$

with g(x) periodic of period 2π. With this in mind, let us write

$$u(x, t) = \sum_n u_n(t)e^{inx}, \tag{3}$$

and substitute into (1). In this way, we obtain an infinite system of nonlinear differential equations

$$u_n'(t) = -n^2 u_n + \sum_{k+1=u} u_k u_1, \qquad u_n(0) = g_n \tag{4}$$

$$n=0,\ \pm 1,\ \pm 2,\ \ldots, \quad \text{where } g(x) \sim \sum_n g_n e^{inx}.$$

In place of studying the partial differential equation of (1), we can investigate the infinite system of (4). It is not a particularly useful method for this purpose, since it appears to require unnecessarily strong restrictions on the Fourier expansion of g as opposed to an alternate treatment of (1) involving integral equations which imposes the conditions directly upon g(x).

The method has, however, become of numerical significance in the last ten years with the development of digital computers capable of obtaining the computational solution of large systems of ordinary differential equations subject to initial conditions. This is consistent with the general idea of using different approaches to an equation to uncover different properties of the solution. To this end, we truncate (4) and consider instead the finite system

$$u_n'(t) = -n^2 u_n + \sum_{k+1=n} u_k u_1, \qquad u_n(0) = g_n \tag{5}$$

$n=0, \pm 1, \ldots, \pm N$, obtained from (4) by supposing that $u_n = 0$ for $|n| > N$. This equation defines for each N a function $u_n^{(N)}(t)$ and an associated function

$$u^{(N)}(x, t) = \sum_{|n| \leq N} u_n^{(N)}(t) e^{inx}.$$

The question we wish to discuss is that of the convergence of $u^{(N)}(x, t)$ to $u(x, t)$ as $N < \infty$. As we shall show, this can be answered in terms of the stability of solution of the equation in (1). It turns out that a slight modification of the foregoing truncation scheme yields a very much simpler approach to the question of convergence.

13. ASSOCIATED EQUATION

Let us introduce the following notation. If

$$f(x) \sim \sum_n f_n e^{inx}, \tag{1}$$

then

$$s_N(f) = \sum_{|n| \leq N} \tag{2}$$

This is to say, s_N is the linear operation which produces the N-th partial sum of the Fourier series for f.

Returning to (12.5), we can write

$$\sum_{|n| \leq N} u_n'(t) e^{inx} = \sum_{|n| \leq N} n^2 u_n e^{inx} + \sum_{|n| \leq N} \sum_{k+\ell=n} u_k u_\ell e^{inx} \tag{3}$$

Let, as in (12.6)

$$u^{(N)}(x, t) = \sum_{|n|<N} u_n(t)\, e^{inx}, \tag{4}$$

suppressing the superscript (N) in $u_n(t)$. Then, in terms of the operation s_N, we may write

$$\frac{\partial u^{(N)}}{\partial t} = \frac{\partial^2 u^{(N)}}{\partial x^2} + s_N[(u^{(N)})^2] \tag{5}$$

If we set

$$r_N(f) = f - s_N(f),$$

we see that (5) takes the form

$$\frac{\partial u^{(N)}}{\partial t} = \frac{\partial^2 u^{(N)}}{\partial x^2} + (u^{(N)})^2 - r_N(u^{(N)})^2. \tag{6}$$

$$u^{(N)}(x, 0) = g - r_N(g).$$

We see then that the effect of the truncation is to alter the original equation of (12.1) by the introduction of a presumably small forcing term, $-r_{(N)}(u^2)$, and by a perturbation of the initial condition, g being replaced by $g - r_{(N)}(g)$.

14. DISCUSSION OF CONVERGENCE OF $u^{(N)}$

The equation

$$\frac{\partial v}{\partial t} = \frac{\partial^2 v}{\partial x^3} + v^2 - r_N(v^2), \qquad (1)$$

$$v(x, 0) = g - r_N(g),$$

cannot be treated in as simple or direct a fashion because of the behavior of the operator s_N. As we know from the theory of Fourier series, there is no relation of the form

$$\|s_N(f)\| \le c_1\|f\|, \quad c_1 > 0, \qquad (2)$$

where $\|f\| = \max |f|$, if we merely postulate continuity of f. To obtain an estimate to the form appearing in (2) and the further estimate

$$\|r_N(f)\| \le \varepsilon\|f\|, \qquad N \ge N_0(\varepsilon)$$

we require some additional assumptions concerning f. To establish these as we go along, employing successive approximations in the treatment of (1), introduces numerous complications. Let us then introduce a slight modification of the original truncation which avoids these difficulties which are extraneous to our original aim of providing a simple computational approach based on the solution of ordinary differential equations.

15. THE FEJER SUM

We begin with the classical result of Fejer. If we use the operation

$$\sigma_N(f) = \frac{s_0 + s_1 + \dots, + s_N}{N+1} = \sum_{|k| \le N} \left(1 - \frac{|k|}{N+1}\right) f_k e^{ikx}, \qquad (1)$$

the (c, 1)-sum, where $f \sim \sum_k f_k e^{ikx}$, we know that

$$\|\sigma(f)\| \le \|f\| \qquad (2)$$

for N=0, 1, ... Furthermore $\sigma_N(f)$ converges uniformly to f as $N \to \infty$.

Let us then replace (14.1) by

$$\frac{\partial w}{\partial t} = \frac{\partial^2 w}{\partial x^2} + \sigma_N(w^2),$$
$$w(x, 0) = \sigma_N(g). \tag{3}$$

It is now easy to employ the standard technique of successive approximation to establish the fact that

$$||w-u|| \le \varepsilon \tag{4}$$

for $N \ge \varepsilon$ under the condition that g is continuous in x.

The proof depends critically upon the fact that the Green's function of the linear equation

$$\frac{\partial u}{\partial t} = \frac{\partial^2 u}{\partial x^2}, \tag{5}$$

is nonnegative, or equivalently upon the monotonicity of the linear partial differential operator.

16. THE MODIFIED TRUNCATION

Writing

$$w(x, t) \quad \sum_{|n| \le N} w_n(t) e^{inx} \tag{1}$$

and substituting in (15.1), we obtain the finite system

$$w_n'(t) = -n^2 w_n(t) + \left(1 - \frac{|n|}{N+1}\right)\left(\sum_{k+\ell=n} w_k w_\ell\right) \tag{2}$$

$n = 0, \pm 1, \pm 2, \ldots, \pm N$, with

$$w_n(0) = \left(1 - \frac{|n|}{N+1}\right) g_n. \tag{3}$$

If we employ this system in place of (12.1), we have a simple proof of the validity of the truncation method. The extension of the foregoing method to numerous classes of partial differential equations is immediate.

BIBLIOGRAPHY AND COMMENTS

Sections 1-11 are largely based on the paper:

Bellman, R., S.P. Azen, and J.M. Richardson, 'On New and Direct Computational Approaches to Some Mathematical Models of Turbulence', Quarterly of Applied Mathematics XXIII (1965), 55-67.

By means of Carleman linearization we can make the infinite system of equations linear. Infinite systems of linear differential equations are studied in:

Bellman, R., Methods of Nonlinear Analysis, Vol. II, Academic Press, Inc., New York, 1973.

Section 12. Many interesting questions of stability arise from truncation. Some of them are studied in:

Bellman, R., 'On the Validity of Truncation for Infinite Systems of Ordinary Differential Equations Associated with Nonlinear Partial Differential Equations', Journal of Mathematical and Physical Sciences 1 (1967) 95-100.

Chapter XV

GREEN'S FUNCTIONS

1. INTRODUCTION

The fascinating concept of the *Green's function* is due to the insight and intuition of George Green (1793-1841), an English mathematician, whose original work was unappreciated for nearly all of his life - largely due to his unusual methodology. George Green was born in 1793 and was 44 years old when he received his degree of Bachelor of Arts in 1837 at Cambridge. His age and his inability to submit to the "systematic training" worked against him and few seemed to realize his remarkable ability at the time. Two years later in 1839 he was elected a Fellow of the College but left almost immediately because of poor health and returned to his home where he died by 1841, a heavy loss indeed to the world of mathematics. His first paper had been published in 1828, five years before beginning formal study. He was entirely self-taught and faced great difficulty in publishing his work.

It is interesting to note that in the following generation, the English physicist, Oliver Heaviside (1850-1925) also had difficulty getting work even accepted for publication for exactly the same reason - unusual methods of solving problems. There is little doubt that examination of records of granting agencies would show such failures were by no means limited to the nineteenth century. There is, however, one valid justification for the lack of appreciation and of widespread use of Green's functions - they are difficult to obtain and consequently are poorly understood and utilized. A student interested in possible use of Green's function techniques can not easily determine if the technique helps with his problem because he is required to read too much formal material. Green's functions have been considered in the realm of rather abstract and formal analysis and generally ignored in course work and books. In principle, Green's functions provide a universal method of solving linear ordinary differential equations. We will show their use can be very significantly extended as part of a general methodology for the solution of *operator equations* which may include *nonlinear stochastic* multidimensional differential equations. We will also simplify the matter of their computation.

2. THE CONCEPT OF THE GREEN'S FUNCTION

Consider a linear differential equation

$$Ly = x. \tag{1}$$

We suppose L is a linear ordinary differential operator - e.g., d/dt or $\alpha(d^2/dt^2) + \beta(d/dt) + \gamma$ - acting on a space of functions $y(t)$, and x is a given function $x(t)$, t being defined on a suitable interval T which may be finite or the entire set of real numbers $[0, \infty)$. We suppose L is invertible, i.e., we can determine the operator L^{-1} which is inverse to L. Then, of course, $L^{-1}L = LL^{-1} = I$ where I is the identity operator. As a simple example, if $L = d/dt$, $L^{-1} = \int dt$. More explicitly, we have $Ly = dy/dt = x(t)$ and $y = \int x(t)dt$. Now the differential equation $Ly = x$ is solving by writing

$$y = L^{-1}x \tag{2}$$

as the desired solution to (1).

Suppose now that L is a more general differential operator than the simple derivative above. It is reasonable to assume L^{-1} is again an integral operator but now involving a kernel $G(t, \tau)$ which by great insight we will call the "Green's function". Thus we write

$$y(t) = L^{-1}x(t) = \int G(t, \tau)x(\tau)\, d\tau \tag{3}$$

We observe that if L is the simple differential operator d/dt, $G(t, \tau)$ must be unity - more specifically it is $H(t-\tau)$ which is the Heaviside function (from Oliver Heaviside mentioned earlier) defined as unity if $t > \tau$ and zero otherwise. This will become very clear later.

Suppose now we operate on both sides of (3) with L again. Then

$$Ly = x(t) = L\int G(t, \tau)x(\tau)\, d\tau \tag{4}$$

L is a differential operator depending on t - such as $L = d/dt + \alpha(t)$. It is correct to write L inside the integral or

$$\int LG(t, \tau)x(\tau)d\tau = x(t) \tag{5}$$

which requires that

$$LG(t, \tau) = \delta(t-\tau) \tag{6}$$

where $\delta(t-\tau)$ is the Dirac delta function. This means that if we replace the forcing function $x(t)$ in (1) with the delta function, the solution will be the Green's function. The remaining mathematicians who will still object to such equations are best simply ignored. The works of Schwartz, Gelfand, Vilenkin, and innumerable others has certainly well justified the brilliant intuitive work of Dirac. It was always correctly used by the physicists but sometimes explained very poorly in standard texts in physics causing controversy. Such equations are understood in the sense of linear functionals and symbolic functions, i.e., in the sense of the theory of distributions.

If we consider the differential equation $Ly(t) = x(t)$ on $[a, b]$, we have seen that the Green's function $G(t, \tau)$ for L satisfies $LG = \delta(t-\tau)$. Hence, G satisfies the homogeneous equation $LG = 0$ if $t \neq \tau$. Suppose we construct a function G such that we can write the solution of the inhomogeneous differential equation $Ly = x(t)$ for $t \in [a, b]$ as

$$y(t) = \int_a^b G(t, \tau)x(\tau)\, d\tau \tag{7}$$

Suppose further that we have determined the solutions $\phi_1, \phi_2, \ldots$ of the homogeneous equation which satisfy the given boundary conditions. Then G satisfies the same equation and boundary conditions, thus the boundary conditions are included in the Green's function. We can use these facts to determine the Green's function after one brief digression to clarify the significance of the two variables t, τ.

Imagine an elastic string along the x axis in the segment $[0, \ell]$. A downward unit force is then applied at the point ξ between 0 and ℓ with its ends fixed. Of course, a real load is applied over a finite area or domain even though it may be small. However, imagine this domain shrinking until in the limit, the load is applied at the point ξ. The string is now displaced at all points except the fixed ends. The elastic deflection at any point x caused by the load at the point ξ is indicated by $G(x-\xi)$, the response at x due to the unit force at ξ. G could therefore be called a response function or an influence function. Now, if an applied force is distributed as $f(x)$, we can imagine it to be decomposed into impulses $\delta(x-\xi)$ with magnitudes $f(\xi)$. Then if we multiply $G(x, \xi)$ by $f(\xi)$ and integrate over all possible ξ, we get the total displacement of the system at the point x, solving the inhomogeneous problem.

Green's function can similarly represent the response at time t to an impulsive force represented by the delta function at time τ. Then it makes sense that if a continuously distributed input $x(t)$ to a dynamical system is decomposed into impulses of magnitude $x(\tau)$ determined by the value of $x(t)$ at $t = \tau$ represented therefore by $x(\tau)\delta(t-\tau)$ - a process not fundamentally different from decomposing $x(t)$ into a Fourier series - then (remembering that $G(t, \tau)$ is the response at time t to an impulse at time τ) it is clear that summing, or actually integrating over all impulses, i.e., over all τ by writing

$$y(t) = \int_{-\infty}^{\infty} G(t, \tau)x(\tau)\, d\tau$$

yields the total response $y(t)$ to the input $x(t)$. Thus G depends on two variables x, ξ in the case of the beam and t, τ in the last example.

We have seen that the superposition principle allows decomposition of any input (forcing function, excitation, stimulus, etc.) into elementary inputs for which the response is known. We can then sum the individual outputs or responses to determine the total response. The expression

$$f(t) = \int_{-\infty}^{\infty} f(\tau)\delta(t-\tau)\, d\tau$$

represents an expansion of a function $f(t)$ into a sequence of impulses.

The Green's function $G(t, \tau)$ is simply the response for such an elementary input - a unit impulse occuring at time $t = \tau$. Then $f(\tau)G(t, \tau)$ is the response for an impulse of magnitude $f(\tau)$ occurring at τ, i.e., $f(\tau)\delta(t-\tau)$. Finally then $\int_{-\infty}^{\infty} G(t, \tau)f(\tau)d\tau$ is the total response for all τ. Since we are interested in real or physical systems, $G(t, \tau)$ must be zero for $t < \tau$. Obviously the response cannot precede the stimulus. Hence the upper limit can be t instead of infinity. The lower limit can be taken as $t = 0$. The initial condition at $t = 0$ takes care of the interval $(-\infty, 0)$. Consequently

$$y(t) = \int_0^t G(t, \tau)x(\tau)\, d\tau$$

We can now consider a differential equation $Lu = f(x)$ where L is a differential operator such as d^2/dx^2. We consider a set or space of functions $u(x)$ which are defined and continuous in the interval $0 \leq x \leq \ell$ and which satisfy $u(0) = u(\ell) = 0$. Let $f(x)$ be an arbitrary (given) continuous function on the same interval. We seek to find u such that $Lu = f(x)$. Our objective is to write $u = L^{-1}f(x)$ as the solution in terms of the *inverse* operator L^{-1} defined by $L^{-1}L = I$, where I is the identity operator. Since L is a differential operator, L^{-1} is clearly an integral operator. Its kernel is called the *Green's function*. This concept is a very useful one for solving inhomogeneous linear differential equations. This integral representation is much more useful for numerical analysis than the differential equation.

The Green's function for a second order equation such as we are considering is a two-point function, e.g., $G(t, \tau)$ or $G(x, \xi)$. It depends only on the differential operator L and the boundary conditions and is independent of the forcing function f. Thus if G is determined, (and the resulting integral exists) we have the solution of the differential equation for any f. The question therefore is: given a differential equation and boundary conditions, can we construct G?

Continuing now with the properties and consequent determination of the Green's function let's consider $a \leq x \leq b$ with ξ an arbitrary point in this interval. (The point ξ will have the significance, of course, of the impulsive point ξ in the first example, or τ in the second example where the interval $[a, b]$ is a time interval.)

3. STURM-LIOUVILLE OPERATOR

The inhomogeneous Sturm-Liouville equation is a linear second order differential equation with variable coefficients - a general type that arises in many applications in physics. It represents an important general example for operator equations of the form

$$Ly(x) = f(x). \quad (1)$$

For the Sturm-Liouville equation, the operator L is defined by

$$L = -(d/dx)[p(x)d/dx] + q(x) \quad (2)$$

along with appropriate boundary conditions satisfied by y(x) at the endpoints a, b of the interval [a, b], i.e., on $a \leq x \leq b$. Thus we are considering the equation

$$-(pu')' + qu = f(x). \quad (3)$$

We assume p is a continuous function and nonzero in [a, b] and that p' and q are continuous in [a, b]. The solution to (1) (or (3)) then is written formally as
$y(x) = L^{-1}f(x) = \int G(x, t)f(t)\, dt$ in terms of the Green's function.

4. PROPERTIES OF THE GREEN'S FUNCTION FOR THE STURM-LIOUVILLE EQUATION

We know from the earlier discussion that the Green's function $G(x, \xi)$ must satisfy the differential equation when the forcing function is replaced by the δ function so that $LG(x, \xi) = \delta(x-\xi)$, i.e.,

$$-[p(x) \frac{dG(x, \xi)}{dx}]' + q(x)G(x, \xi) = \delta(x-\xi). \quad (1)$$

Hence G satisfies the homogeneous equation LG = 0 if $x \neq \xi$.

Now, suppose we can construct a function $G(x, \xi)$ such that we can write the solution of the inhomogeneous Sturm-Liouville differential equation $Lu(x) = f(x)$ as

$$u(x) = \int_a^b G(x, \xi)f(\xi)\, d\xi \quad (2)$$

and we have the solutions of the *homogeneous* Sturm-Liouville equation $u_1(x)$ satisfying the boundary condition at x = a and the solution $u_2(x)$ satisfying the boundary conditions at x = b, i.e., we have solutions of Ly = 0 in the two regions $x < \xi$ and $x > \xi$.

We can indicate the appropriate region of validity by simply identifying $G(x, \xi)$ for $x < \varepsilon$ as G_1, and correspondingly $G(x, \xi)$ for $x > \xi$ as G_2. Thus for a given source point ξ we speak of the Green's function G_1 for $x < \xi$ and G_2 for $x > \xi$ with the following required properties:

(1) The functions G_1 and G_2 each satisfy the homogeneous Sturm-Liouville equations $LG = 0$ in their respective intervals of definition hence $LG_1 = 0$ for $x < \xi$ and $LG_2 = 0$ for $x > \xi$. Hence we write $G_1 = c_1u_1$ and $G_2 = c_2u_2$.

(2) G satisfies the homogeneous boundary conditions at $x = a$ and at $x = b$; i.e., G_1 satisfies the boundary conditions we impose on the solution of the differential equation $Lu = f(x)$ at $x = a$, and G_2 satisfies the boundary conditions at $x = b$, e.g., $u(a) = u'(a) = 0$ and $u(b) = u'(b) = 0$.

(3) G is continuous at $x = \xi$, i.e., $G_1 = G_2$.

Now consider the behavior of G for x very close to the source point ξ. We integrate the equation (1) from $\xi - \varepsilon$ to $\xi + \varepsilon$; this yields

$$-p(x)\,\frac{dG(x, \xi)}{dx}\Bigg|_{\xi-\varepsilon}^{\xi+\varepsilon} + \int_{\xi-\varepsilon}^{\xi+\varepsilon} q(x)G(x, \xi)$$
$$= -p(\xi+\varepsilon)\frac{dG(\xi+\varepsilon, \xi)}{dx} + p(\xi-\varepsilon)\frac{dG(\xi-\varepsilon, \xi)}{dx} + \int_{\xi-\varepsilon}^{\xi+\varepsilon} q(x)G(x, \xi) = 1 \tag{4}$$

The last integral goes to zero as $\xi \to 0$ since q and G are continuous. In the limit then as $\varepsilon \to 0$ we have

$$-p(\xi)\,\frac{dG(\xi+0, \xi)}{dx} + p(\xi)\,\frac{dG(\xi-0, \xi)}{dx} = 1$$

or,

$$-p(\xi)\left[\frac{dG(\xi+0, \xi)}{dx} - \frac{dG(\xi-0, \xi)}{dx}\right] = 1. \tag{5}$$

We see then an important property of the Green's function. There exists a step discontinuity or jump in the derivative of G with respect to x at the point $x = \xi$ and its magnitude must be $-1/p(\xi)$. This discontinuity must be finite - we initially 'cleverly' assumed p to be non-zero in the interval. Continuity of G at ξ requires $c_2u_2 - c_1u_1 = 0$. Now the discontinuity of the first derivative of G results in

$$c_2u_2'(\xi) - c_1u_1(\xi) = -1/p(\xi) \tag{6}$$

We have therefore two equations to solve for c_1 and c_2. A unique solution exists provided

$$\begin{vmatrix} u_1(\xi) & u_2(\xi) \\ u_1'(\xi) & u_2'(\xi) \end{vmatrix} = u_1(\xi)u_2'(\xi) - u_2(\xi)u_1'(\xi) \neq 0 \tag{7}$$

This determinant is called the *Wronskian* of the solutions u_1 and u_2 of the homogeneous equation. The non-vanishing of the Wronskian is a necessary condition for linear independence of the functions u_1 and u_2.

We observe that $G(x, \xi)$ satisfies the homogeneous equation except at $x = \xi$. It is clear also that $G(x, \xi)$, and consequently the solution of $Lu = f$, is continuous at $x = \xi$. Finally G must satisfy the boundary conditions - it depends both on the differential operator L and on the boundary conditions that $u(x)$ must satisfy.

Identifying

$$c_1 = -u_2(\xi)/pW$$

$$c_2 = -u_1(\xi)/pW$$

Now

$$G = \begin{cases} c_1u_1(x) = -u_2(\xi)u_1(x)/pW & \text{if } x < \xi \\ c_2u_2(x) = -u_1(\xi)u_2(x)/pW & \text{if } x > \xi \end{cases} \tag{8}$$

Equivalently, we can write the convenient form

$$G(x, \xi) = \frac{u_1(x)u_2(\xi)H(\xi-x) + u_2(x)u_1(\xi)H(x-\xi)}{pW} \tag{9}$$

where W is the Wronskian and $H(\xi-x)$ is the Heaviside step function which is unity for $\xi \geq x$ and zero otherwise.

We can simply state the properties we found to be

necessary as prior conditions and derive the Green's function, an approach appealing to mathematicians and often frustrating to others.

(i) The interval $a \leq x \leq b$, or $[a, b]$, is divided into two regions by a parameter we will call ξ. Let $G(x) = G_1$ for the first or left region specified by $a \leq x < \xi$;

(ii) let G_1, G_2 satisfy the *homogeneous* Sturm-Liouville equation. (We have seen that G satisfies the homogeneous equation at points other than ξ.) Consequently we have

$$
\begin{aligned}
LG_1(x) &= 0 \qquad a \leq x < \xi \\
LG_2(x) &= 0 \qquad \xi < x \leq b;
\end{aligned}
\tag{10}
$$

(iii) let G_1 satisfy the boundary conditions imposed on $u(x)$ at $x = a$ and let G_2 satisfy the boundary conditions imposed on $u(x)$ at $x = b$;

(iv) assume $u(a) = u'(a) = 0$ and $u(b) = u'(b) = 0$ are the given boundary conditions;

(v) require that $G_1(\xi) = G_2(\xi)$, i.e., G is continuous at the boundary point between the two regions;

(vi) finally, require that $G'(x)$ be discontinuous at ξ and specifically that

$$
\left.\frac{dG_2(x)}{dx}\right|_{\xi} - \left.\frac{dG_1(x)}{dx}\right|_{\xi} = -\frac{1}{p(\xi)}
\tag{11}
$$

where the p comes from the operator L.

We see that $G = G(x, \xi)$ depends on both the operator and the boundary conditions. To construct G, let $v(x)$ be a solution of the homogeneous Sturm-Liouville equation satisfying the boundary conditions at $x = a$ and let $w(x)$ be a solution of the same equation satisfying the boundary conditions at $x = b$. (We know, of course, there should be two linearly independent solutions, say $v(x)$ and $w(x)$. Then $c_1 v(x) + c_2 w(x) = 0$ cannot be satisfied unless both c_1 and c_2 are identically zero.) Then, take

$$
\begin{aligned}
G(x, \xi) &= c_1 v(x) \quad \text{for} \quad a \leq x < \xi \\
G(x, \xi) &= c_2 w(x) \quad \text{for} \quad \xi < x \leq b
\end{aligned}
\tag{12}
$$

Because of the continuity at ξ (condition v)

$$c_2 w(\xi) - c_1 v(\xi) = 0 \tag{13}$$

Again, because of condition (vi),

$$c_2 w'(\xi) - c_1 v'(\xi) = -1/p(\xi). \tag{14}$$

We get a unique solution for c_1, c_2 only if v, w are linearly independent. The condition for linear independence can be given by stating that the Wronskian W is non-zero. It is in fact given by

$$W = \begin{vmatrix} v(t) & w(t) \\ v'(t) & w'(t) \end{vmatrix} = A/p \tag{15}$$

where A is a constant (Abel's constant). This means the following: If v(x) and w(x) are linearly independent for all x in $a \le x \le b$, then the equation

$$c_1 v(x) + c_2 w(x) = 0 \tag{16}$$

cannot be satisfied for all $x \in [a, b]$ unless $c_1 = 0$ and $c_2 = 0$. (In terms of vectors, w doesn't lie along the direction of v.) Suppose we say v, w are linearly dependent then $c_1 v + c_2 w = 0$ for nonzero c_1, c_2. Differentiating, we can write $c_1 v' + c_2 w' = 0$ and consider this as two equations in the unknown c_1, c_2. A nontrivial solution exists if

$$\begin{vmatrix} v & w \\ v' & w' \end{vmatrix} = 0$$

i.e., linear dependence implies vanishing of the Wronskian. (The converse is not true.) A precise statement is given by the following well-known theorem. The necessary and sufficient condition for v(x) and w(x) to be linearly dependent in [a, b] is that their Wronskian is zero for $x \in [a, b]$ provided v, w are solutions of $y'' + p(x)y' + q(x)y = 0$ in which p, q are continuous on [a, b].

It is easy now to prove this Green's function solves the inhomogeneous equation.

Proof:

$$u(x) = -\frac{1}{pW}\int_a^x w(x)v(t)f(t)\,dt - \frac{1}{pW}\int_x^b v(x)w(t)f(t)\,dt$$

$$u'(x) = -\frac{1}{pW}\int_a^x w'(x)v(t)f(t)\,dt - \frac{1}{pW}\int_x^b v'(x)w(t)f(t)\,dt$$

(noting that derivatives of limits cancel out)

$$u''(x) = -\frac{1}{pW}\int_a^x w''(x)v(t)f(t)\,dt - \frac{1}{pW}\int_x^b v''(x)w(t)f(t)\,dt$$

$$-\frac{1}{pW}[v(x)w'(x) - w(x)v'(x)]f(x)\,.$$

Hence from (15)

$$u''(x) = -\frac{w''(x)}{pW}\int_a^x v(t)f(t)\,dt - \frac{v''(x)}{pW}\int_x^b w(t)f(t)\,dt$$

$$- f(x)/p(x)$$

Substituting into the Sturm-Liouville equation, we have

$$Lu(x) = -\frac{[Lw(x)]}{pW}\int_a^x v(t)f(t)\,dt$$

$$-\frac{[Lv(x)]}{pW}\int_a^b w(t)f(t)\,dt - f(x)\,.$$

We observe immediately that the first two terms vanish since v, w are chosen to satisfy the homogeneous Sturm-Louiville equation.

Now verify that the solution u(x) satisfies the boundary conditions. At x = a,

$$u(a) = -\frac{v(a)}{pW}\int_a^b w(t)f(t)\,dt = cv(a) = 0$$

since the definite integral is merely a constant, and

$$u'(a) = -\frac{v'(a)}{pW}\int_a^b w(t)f(t)\,dt = cv'(a) = 0$$

Analogous results are easily verified at u = b. Hence

$$u(x) = \int_a^b G(x, t)f(t)\,dt$$

satisfies

$$Lu(x) + f(x) = 0$$

and the boundary conditions, i.e., *the boundary conditions are built into the Green's function.*

5. PROPERTIES OF THE δ FUNCTION

The most important property of the δ function is that for every continuous function ϕ

$$\int_{-\infty}^{\infty}\delta(x)\phi(x)\,dx = \phi(0) \tag{1}$$

i.e., the function picks out the value of $\phi(x)$ at the point where the argument of the δ function is zero. Thus $\delta(x-x')$ in the above integral would result in the value $\phi(x')$.

The δ function can be handled algebraically like an ordinary function if we interpret an equation involving $\delta(x)$ in the following way. If the equation is multiplied throughout by an arbitrary continuous function $\phi(x)$ (a function of compact support) and integrated from $-\infty$ to ∞, then using (1) to evaluate integrals, the resulting equation involves only ordinary functions and is correct. We can, for example, write $x\delta(x) = 0$ because

$$\int_{-\infty}^{\infty}\delta(x)x\phi(x)\,dx = \int_{-\infty}^{\infty}\delta(x)\Psi(x)\,dx = \Psi(0) \text{ which equals zero since}$$

$\Psi(x) = x\delta(x)$.

Exercise: Show $\delta(x) = \delta(-x)$.

Exercise: Show $\int_{-\infty}^{\infty} \delta(ax-b)\phi(x)\,dx = a^{-1}\phi(ba^{-1})$.

The function $\phi(x)$ is called a *testing function*. It's convenient to restrict the functions $\phi(x)$ to being continuous, with continuous derivatives of all orders, and vanishing outside a finite interval. The latter requirement restriction is called *compact support*. A space (set of functions) satisfying these requirements will be denoted by C_0^∞. (It is a linear vector space.)

Some further properties of δ-functions:

$$\delta(ax) = \frac{1}{|a|}\delta(x) \quad \text{(a is real and non-zero)}$$

$$\delta(f(x)) = \sum_k \frac{\delta(x-x_k)}{|f'(x_k)|} \quad (x_k \text{ are zeros of } f(x) \text{ and } f'(x) \text{ continuous at } x = x_k)$$

$$\delta'(x) = -\frac{1}{x}\,\delta(x)$$

$$\delta^{(k)}(x) = (-1)^k \frac{k!}{x^k}\,\delta(x)$$

$$\delta^{(k)}(-x) = (-1)\,\delta^{(k)}(x)$$

6. DISTRIBUTIONS

A *linear functional* $F(\phi)$ on the space C_0^∞ of testing functions $\{\phi(x)\}$ is defined by assigning a real or complex number $F(\phi)$ to every ϕ such that

$$F(\phi_1 + \phi_2) = F(\phi_1) + F(\phi_2)$$

$$F(k\phi) = k\,F(\phi)$$

for a scalar k. A sequence of testing functions $\{\phi_n\}$ converges to zero if all ϕ_n vanish outside the same finite interval of support, if the functions ϕ_n and all their derivatives converge uniformly to zero. $F(\phi)$ is continuous if the sequence of numbers $F(\phi_n)$ converges to zero whenever the sequence $\{\phi_n\}$ converges to zero. Such a continuous linear functional is a *distribution* (in the sense of Schwartz) or a *generalized function*. Examples are $F(\phi) = \phi'(0)$, or

$$F(\phi) = \int_0^1 \phi(x)\, dx.$$

7. SYMBOLIC FUNCTIONS

The continuous linear functional $F(\phi)$ is expressed as an integral on a space of testing functions $\phi(x)$ vanishing outside a finite interval

$$F(\phi) = \int_{-\infty}^{\infty} s(x)\phi(x)\, dx$$

where $s(x)$ is a *symbolic function*, a generalization including ordinary functions $f(x)$. However, $s(x)$ need not have numerical values *unless* multiplied by a testing function and integrated. Such a *symbolic function* is not a function in the usual sense. $\delta(x)$, $\delta'(x)$, $H(x)$ are examples. Actually one realizes that many functions of physics, such as temperature or density for example, make physical sense when integrated over a small region. Temperature or density at a point makes little sense so such integrals or functionals are more fundamental in nature.

8. DERIVATIVE OF SYMBOLIC FUNCTIONS

It is necessary to define the derivative of a symbolic function to be consistent with an ordinary derivative if the symbolic function is an ordinary function. If $f(x)$ is an ordinary integrable function whose derivative is $f'(x)$ then

$$\int_{-\infty}^{\infty} f'(x)\phi(x)\, dx = -\int_{-\infty}^{\infty} f(x)\phi'(x)\, dx$$

using integration by parts and the fact that $\phi(x)$ vanishes outside a finite interval (i.e., ϕ has compact support). Now if

$$\int_{-\infty}^{\infty} s'(x)\phi(x)\, dx = -\int_{-\infty}^{\infty} s(x)\phi'(x)\, dx$$

for every testing function $\phi(x)$ then we say $s'(x)$ is the derivative of the symbolic function $s(x)$. Thus our definition (1) of the δ function can be written as

$$F(\phi) = \phi(0) = \int_{-\infty}^{\infty} \delta(x)\phi(x)\, dx$$

Now the derivative $\delta'(x)$ of the δ function is defined by

$$\int_{-\infty}^{\infty} \delta'(x)\phi(x)\, dx = - \int_{-\infty}^{\infty} \delta(x)\phi'(x)\, dx = -\phi'(0)$$

Thus δ' produces a functional which assigns the value $-\phi'(0)$ to $\phi(x)$. Similarly δ'' is the derivative of δ' and assigns the value $\phi''(0)$ to $\phi(x)$.

The *Heaviside function* $H(x)$ is defined as 1 for $x \geq 0$ and 0 for $x < 0$. Its derivative can now be obtained by writing:

$$\int_{-\infty}^{\infty} H'(x)\phi(x)\, dx = - \int_{-\infty}^{\infty} H(x)\phi'(x)\, dx$$

$$= - \int_{0}^{\infty} \phi'(x)\, dx = \phi(0)$$

Noting again that $\int_{-\infty}^{\infty} \delta(x)\phi(x)\, dx = \phi(0)$ then

$$H'(x) = \delta(x)$$

i.e., the δ function is the derivative of the Heaviside function. Now the notion of derivative has been generalized to the derivative of a function $H(x)$ which has a jump discontinuity at $x = \xi$ of magnitude 1. Suppose the jump has magnitude m. We can represent it with $mH(x-\xi)$. Writing

$$f(x) = g(x) + mH(x-\xi)$$

where f, g are differentiable in the ordinary sense,

$$\int f(x)\phi'(x)\,dx = \int [g(x) + mH(\xi - x)]\phi'(x)\,dx$$

$$= \int g(x)\phi'(x)\,dx + m\int H(x-\xi)\phi'(x)\,dx$$

$$= -\int g'(x)\phi(x)\,dx - m\phi(\xi)$$

but

$$\int f'(x)\phi(x)\,dx = -\int f(x)\phi'(x)\,dx$$

$$= \int g'(x)\phi(x)\,dx + m\int \delta(x-\xi)\phi(x)\,dx$$

i.e.,

$$f'(x) = g'(x) + m\delta(x-\xi)$$

in the sense of distributions (multiplying by a testing function and integrating. All integrals are over the entire real line $-\infty$ to ∞.

Example: Consider the motion (velocity) of a particle of mass m in a resistive medium under the influence of an external force f(t). We write

$$m\,dv/dt + Rv = f(t) \tag{1}$$

The Green's function for the velocity satisfies

$$m dG(t, \tau)/dt + RG(t, \tau) = \delta(t-\tau) \tag{2}$$

$G(t, \tau) = G(t-\tau)$ since the system is time-invariant but it will work out anyway. For $t < \tau$, there is no motion. For $t > \tau$, the velocity or G is given by $(1/m)\exp\{-(R/m)(t-\tau)\}$. Thus G is the response in this case - the motion or velocity - resulting from a unit impulse $\delta(t-\tau)$ delivered at time τ. At time τ, the velocity because of the unit impulse becomes 1/m

$$G(t, \tau) = (1/m)\exp\{-(R/m)(t-\tau)\} \quad \text{for } t > \tau \tag{3}$$

or in terms of the Heaviside function or step function $H(t-\tau)$ defined to be 1 for $t \geq \tau$ and zero for $t < \tau$ we can write

$$G(t, \tau) = \frac{H(t-\tau)}{m} \exp\{-(R/m)(t-\tau)\} \tag{4}$$

The solution to the differential equation is therefore

$$v(t) = (1/m) \int_0^t H(t-\tau) \exp\{-(R/m)(t-\tau)\} f(\tau)\, d\tau \tag{5}$$

which can also be found using an integrating factor on the first order equation for v.

Consider now the operator $L = d^2/dx^2$ and the equation $Lu(x) = f(x)$. The Green's function satisfies $LG(x, \xi) = \delta(x-\xi)$. Therefore

$$\frac{d^2G}{dx^2} = \delta(x-\xi)$$

Integrating,

$$\frac{dG}{dx} = H(x-\xi) + \alpha(\xi)$$

where α is an arbitrary function. Integrating again

$$G(x, \xi) = \int H(x-\xi)\, dx + x\alpha(\xi) + \beta(\xi)$$

$$= (x-\xi)H(x-\xi) + x\alpha(\xi) + \beta(\xi) \tag{6}$$

G is a continuous piece-wise differentiable function and so long as $f(x)$ is an integrable function vanishing outside a finite interval, the solution of the differential equation is

$$u(x) = \int G(x,\xi)f(\xi)\, d\xi \tag{7}$$

where α, β must be chosen to satisfy the boundary conditions. Since this satisfies $Lu = f$,

$$Lu(x) = f(x) = L\int G(x,\xi)f(\xi)\, d\xi$$

$$= \int LG(x,\xi)f(\xi)\, d\xi$$

$$= \int \delta(x-\xi)f(\xi)\, d\xi$$

so that $LG(x,\xi) = \delta(x-\xi)$ as we have said.

The arbitrary functions α, β depend on the assumed initial or boundary conditions that u must satisfy. G must satisfy the same conditions so we can impose the conditions either on (6) or on (7). Suppose we are given two-point boundary conditions $u(0) = u(1) = 0$. Then in (6), $G(0, \xi) = G(1, \xi) = 0$ and we find immediately $\beta = 0$ and $\alpha = -(1-\xi)$. Hence for x in [0, 1] and for $\varepsilon \leq 1$,

$$G(x, \xi) = (x-\xi)H(x-\xi)-x(1-\xi)$$

Exercise: Impose the boundary conditions on $u(x) = \int G(x,\xi)f(\xi)\, d\xi$ with the given G, i.e.,

$$u(x) = \int_{-\infty}^{x} (x-\xi)f(\xi)\, d\xi + x\int_{-\infty}^{\infty} \alpha(\xi)f(\xi)\, d\xi + \int_{-\infty}^{\infty} \beta(\xi)f(\xi)\, d\xi$$

Show that $\beta = \xi H(-\xi)$ and $\alpha = -1 + \xi H(\xi) = 0$ to obtain the final form for u, identify the kernel as $G(x, \xi) = (x-\xi)H(x-\xi)-x(1-\xi)$, and finally show G satisfies the same boundary conditions.

Example: Show for $L = -d^2/dx^2$ and $u(0) = u'(0) = 0$ that $G(x, \xi) = -(x-\xi)H(x-\xi)$ for all x, and, $u(x) = -\int_0^x f(\xi)(x-\xi)\, d\xi$. We get immediately $G = \alpha x+\beta$ for $x > \xi$. The jump in the derivative of G is -1 hence $\alpha = -1$. Continuity at $x = \xi$ gives $\beta = \xi$. Hence $G = \xi-x$ for $x > \xi$. For $x < \xi$, of course, $G = 0$. Therefore $G(x, \xi) = -(x-\xi)H(x-\xi)$ for all x. The rest is quickly verified.

9. WHAT SPACE ARE WE CONSIDERING?

We are considering linear second-order differential operators L given by $L = a(x)d^2/dx^2 + b(x)d/dx + c(x)$. The coefficients a, b, c are assumed to be continuous functions on a finite interval I. The interval can be $a \leq x \leq b$ for finite a, b or $0 \leq x \leq 1$ for convenience. We can indicate the interval as I(a, b) or [a, b] . Let's denote by S a space of (Lebesgue-integrable) functions u(x) such that $\int_0^1 u^2(x)\, dx < \infty$, i.e., an L_2 space. We will consider functions u for which Lu is defined and in S. In general of course, not all $u \in S$ may be appropriately differentiable. Even if $u \in S$, Lu may not. We are therefore considering L to only act on functions for which $Lu \in S$. The domain of L is the set $S = \{u(x)\}$ of functions with piece-wise continuous second derivatives, which satisfy given boundary conditions and result in $Lu \in S$. (This domain is called a linear manifold.)

Example: Steady heat flow on a homogeneous rod whose ends are at 0 and 1 on the x axis. We can model this with

$$-d^2u/dx^2 = f(x) \qquad 0 < x < 1$$

$$u(0) = \alpha \qquad u(1) = \beta$$

where α, β, the prescribed temperatures of the ends, and f(x), the forcing function or heat source density, are given data (and the product of thermal conductivity and cross-sectional area is unity). We write the usual equation for the Green's function

$$-d^2G(x, \xi)/dx^2 = \delta(x-\xi) \tag{1}$$

Think what this means physically. We have introduced a new source - not distributed - a concentrated source of unit strength at $x = \xi$, a point inside the rod, and zero temperatures at the ends. The source acts as a heat input and the resulting response is the Green's function $G(x, \xi)$. The temperature at the point x due to a source at ξ. G vanishes at x = 0 and x = 1. We find that

$$\begin{aligned} G &= Ax \qquad & 0 < x < \xi \\ G &= B(1-x) \qquad & \xi < x < 1 \end{aligned} \tag{2}$$

A, B are independent of x but can depend on ξ. Notice G and G' cannot both be continuous at $x = \xi$; that would imply A and B both zero. We know of course that G is continuous at $x = \xi$ but G' has a jump discontinuity there. The continuity condition gives us $A\xi = B(1-\xi)$ so $A = B(1-\xi)/\xi$. The derivative conditions gives $-A-B = -1$. We previously found the magnitude of the jump for Sturm-Liouville operators. Here, we have $p=1$. From these conditions on A and B we find $A = 1-\xi$ and $B = \xi$. Consequently

$$G(x, \xi) = (1-\xi)x \qquad 0 \leq x < \xi$$
$$G(x, \xi) = (1-\xi)\xi \qquad \xi < x \leq 1 \tag{3}$$

This problem and other very interesting physical examples appear in Green's Functions and Boundary Value Problems (Ivar Stakgold, Wiley, 1979, Chapter 1).

Example: Consider the Newton's law problem $m\, d^2u/dt^2 = F(t)$ for $t > 0$ for initial conditions $u(0) = \alpha$ and $u'(0) = \beta$. The equation satisfied by the Green's function is $m d^2G/dt^2 = \delta(t-\tau)$ where G yields the position of the particle at time t when subjected to a unit impulse at time τ. Thus $G = 0$ until $t = \tau$. With the continuity and jump conditions, we find that

$$G(t, \tau) = (1/m)(t-\tau)H(t-\tau) \tag{4}$$

Example: $L = -d^2/dx^2$ with $u(a) = u(b) = 0$. The solution of the homogeneous equation $Lu = 0$ is $Ax + B$ where A and B are constants. At $x = a$ we have $u(a) = 0$ hence $Aa + B = 0$ or $B = -Aa$. Therefore we write the first of the two solutions to this homogeneous equation as $u_1(x) = A(x-a)$. Similarly the second solution u_2 satisfies the boundary condition at $x = b$ and hence $Ab + B = 0$ so that $u_2(x) = A(x-b)$. The Wronskian of u_1, u_2 is

$$\begin{vmatrix} A(x-a) & A(x-b) \\ A & A \end{vmatrix} = A^2(b-a)$$

We write for the region $x < \xi$,

$$G_1(x, \xi) = c_1u_1(x)$$

and for the region $x > \xi$,

$$G_2(x, \xi) = c_2u_2(x).$$

Finding c_1, c_2 we arrive at

$$G_1 = -u_2(\xi)u_1(x)/A$$

$$G_2 = -u_1(\xi)u_2(x)/A.$$

The Wronskian of u_1, u_2 is

$$\begin{vmatrix} u_1 & u_2 \\ u_1' & u_2' \end{vmatrix} = u_1u_2' - u_2u_1' = A/p = -A$$

Therefore

$$\begin{aligned} G_1 &= \frac{-u_2(\xi)u_1(x)}{A^2(b-a)} = \frac{-(\xi-b)(x-a)}{b-a} \\ G_2 &= \frac{-u_1(\xi)u_2(x)}{A^2(b-a)} = \frac{-(\xi-a)(x-b)}{b-a} \end{aligned} \tag{5}$$

<u>Example</u>: $d^2u/dx^2 + k^2u = 0$ with $u(0) = u(a) = 0$. We write $u_1 = \sin kx$ which will vanish at $x = 0$ and $u_2 = \sin k(a-x)$ which will vanish at $x = a$. The Wronskian is

$$\begin{aligned} W &= \sin kx[-k \cos k(a-x)] - \sin k(a-x)[k \cos kx] \\ &= -k[\sin kx \cos k(a-x) + \cos kx \sin k(a-x)] \\ &= -k \sin ka \end{aligned}$$

and the Green's function is given by

$$G(x, \xi) = \frac{-\sin k(a-\xi) \sin kx}{k \sin ka} \qquad x < \xi$$

$$G(x, \xi) = \frac{-\sin k(a-x) \sin k\xi}{k \sin ka} \qquad x < \xi \tag{6}$$

(We notice if a is a nonzero integer, G has poles in the complex k plane for $k = n\pi/a$.)

10. BOUNDARY CONDITIONS

We must also require the solution u to satisfy suitable boundary conditions which will be indicated by $B_1(u)$ and $B_2(u)$. These can be

$$B_1(u) = \alpha_{10}u(0) + \alpha_{11}u'(0) + \beta_{10}u(1) + \beta_{11}u'(1) = b_1$$

$$B_2(u) = \alpha_{20}u(0) + \alpha_{21}u'(0) + \beta_{20}u(1) + \beta_{21}u'(1) = b_2$$

where the α's, β's, and b's are constants. The proper choice of the constants specifies the appropriate boundary conditions. If we have a condition such as $au(0) + bu'(0) = 0$ which involve the function u and its derivative u' at either $x = 0$ of $x = 1$, but not both endpoints, this means *unmixed boundary conditions*. They are called *periodic boundary conditions* if they are the same at both boundaries, e.g. $u(0) = u(1)$ or $u'(0) = u'(1)$. Let's consider some examples:

Example 1: Find $G(x, \xi)$ for $L = -d^2/dx^2$ and initial conditions $u(0) = u'(0)$.

$$\frac{d^2G}{dx^2} = -\delta(x-\xi)$$

$$\frac{dG}{dx} = -H(x-\xi) + \alpha(\xi)$$

$$G = -(x-\xi)H(x-\xi) + x\alpha(\xi) + \beta(\xi) .$$

Using the given conditions for G, i.e., $G(0, \xi) = G_x(0, \xi) = 0$ results in $\alpha = \beta = 0$ and consequently

$$G(x, \xi) = -(x-\xi)H(x-\xi)$$

and

$$u(x) = -\int_0^1 (x-\xi)H(x-\xi)f(\xi)\, d\xi = -\int_0^x (x-\xi)f(\xi)\, d\xi$$

We can obtain the same result by a technique we will see to be more useful in problems involving a more complicated operator in which we can't simply integrate twice as above.

Notice $d^2G/dx^2 = 0$ at points other than $x = \xi$. Thus G satisfies the homogeneous equation everywhere except the singularity at $x = \xi$. Solving the homogeneous equation we have $G = \alpha(\xi)x + \beta(\xi)$. Since the boundary conditions require that $G(0, \xi) = G_x(0, \xi) = 0$, $\alpha = \beta = 0$ for $x < \xi$ where G satisfies the homogeneous equation. Now for $x > \xi$, i.e., between $x = \xi$ and $x = 1$, G again equals $\alpha(\xi)x + \beta'(\xi)$. We know we must have continuity of G at $x = \xi$ and that there must be a jump in the derivative of -1. Hence dG/dx changes from zero to -1 at $x = \xi$. Continuity at $x = \xi$ requires

$$\lim_{x\to\xi^-} G(x, \xi) = \lim_{x\to\xi^+} G(x, \xi)$$

which implies $\beta = \xi$ since $G = 0$ for $x < \xi$. We have now

$$G(x, \xi) = 0 \qquad x < \xi$$

$$G(x, \xi) = -x+\xi \qquad x > \xi$$

or simply

$$G(x, \xi) = -(x-\xi)H(x-\xi)$$

for all x.

Example 2. For $L = -d^2/dx^2$, consider initial conditions $u(0) = a$, $u'(0) = b$. Find $G(x, \xi)$ to solve $Lu = f(x)$. We have

$$G(x, \xi) = -(x-\xi)H(x-\xi)-x\alpha(\xi)-\beta(\xi)$$

Evaluating G and G_x at $x = 0$ and imposing the given conditions results in $\alpha = -b$ and $\beta = -a$. Thus

$$G(x, \xi) = a + bx - (x-\xi)H(x-\xi)$$

or

$$u(x) = a + bx \int_0^x (x-\xi)f(\xi)\, d\xi$$

Example 3. With the same operator, consider two-point boundary conditions $u(0) = 0$ and $u(1) = 0$

$$G(x, \xi) = -(x-\xi)H(x-\xi) + x\alpha(\xi) + \beta(\xi)$$

$$\beta = 0$$

$$\alpha = (1-\xi)$$

so that

$$G(x, \xi) = -(x-\xi)H(x-\xi) + (1-\xi)x$$

Thus at $x < \xi$, $G(x, \xi) = x(1-\xi)$. For $x > \xi$, $G = \xi(1-x)$ and we can write the solution u as

$$u(x) = -\int [(x-\xi)H(x-\xi) + (1-\xi)x]f(\xi)\, d\xi$$

$$= -\int_0^x (x-\xi)f(\xi)\, d\xi + x\int_0^x (1-\xi)f(\xi)\, d\xi$$

We can also obtain the solution as follows: in the region to the left of $x = \xi$ and including $x = 0$, G satisfies the homogeneous equation $Lu = 0$ hence

$$d^2G/dx^2 = 0$$

which means G is a linear function of x, i.e., $G = c_1x +$ constant. However, since $G(0, \xi) = 0$ we have simply $G = c_1x$ in this region which we will call region I identifying G in this region as G_I. For the region $\xi < x \leq 1$ denoted as region II, we write $G_{II} = c_2 + c_3x$. Since $G(1, \xi) = 0$, $c_2 + c_3 = 0$ or $c_2 = -c_3$. Hence $G_{II} = c_2(1-x)$ for $\xi < x \leq 1$.

Using the continuity of G at $x = \xi$, $G_I = G_{II}$ at $x = \xi$. Also $dG_{II}/dx - dG_I/dx = -1$. Then $G_I = c_1 x = (1-\xi)x$ and $G_{II} = \xi(1-x)$.

We can look at it in this way. To meet the continuity and derivative requirements, we multiply G_I by the value of G_{II} at $x = \xi$ and vice versa.

Thus

$$
\begin{aligned}
G &= x(1-\xi) \qquad x < \xi \\
&= (1-x)\xi \qquad x > \xi
\end{aligned}
$$

or

$$G = x(1-\xi)H(\xi-x) + (1-x)\xi H(x-\xi)$$

which is the Green's function to solve the equation $Lu = f(x)$ for the solution $u(x)$ satisfying the boundary conditions.

Example 4. For $L = -d^2/dx^2$ and $u(0) = u'(0)$ we have $-d^2G(x, \xi)/dx^2 = \delta(x-\xi)$, $G(0) = G'(0) = 0$. G satisfies $d^2G/dx^2 = 0$ except at $x = \xi$ where G is continuous but G' jumps with a magnitude of -1.

$$G = ax + b$$

Since $G(0) = 0$, $b = 0$. Further $G' = a$ but $G'(0) = 0$ so $a = 0$. Therefore $G = 0$ for $x < \xi$. For $x > \xi$, $G = ax + b$. Since the derivative jumps by -1, $a = -1$ and since G is continuous at $x = \xi$, $b = \xi$ therefore

$$
\begin{aligned}
G(x, \xi) &= 0 \qquad x < \xi \\
G(x, \xi) &= \xi - x \qquad x > \xi
\end{aligned}
\tag{1}
$$

Thus

$$G(x, \xi) = -(x-\xi)H(x-\xi)$$

for all x. Hence

$$u(x) = \int_0^x (x-\xi)H(x-\xi)f(\xi)\, d\xi$$

Example 5: $L = -d^2/dx^2$, $u(0) = u(1) = 0$

$$-d^2G/dx^2 = \delta(x-\xi) \tag{2}$$

$$G(0, \xi) = G(1, \xi) = 0$$

For $x < \xi$, the solution u_1 of the homogeneous equation, also satisfying the boundary condition on the left, $G(0, \xi) = 0$, will be $u_1 = c_1 x$.

For $x > \xi$, the solution u_2 of the homogeneous equation satisfying the right boundary condition $G(1, \xi) = 0$ is $u_2 = c_2(1-x)$.

$$G_1(x, \xi) = (1-\xi)x$$

$$G_2(x, \xi) = \xi(1-x) .$$

The derivative for $x < \xi$ is $1-\xi$. The derivative for $x > \xi$ is $-\xi$. The jump at $x = \xi$ is $-\xi-(1-\xi) = -1$. Hence

$$G_2' - G_1' = -1$$

$$u_1 u_2' - u_2 u_1' = A.$$

Finally,

$$\begin{aligned} G_1 &= (1-\xi)x \\ G_2 &= \xi(1-x) \end{aligned} \tag{3}$$

or $G(x, \xi) = x(1-\xi)H(\xi-x) + (1-x)\,H(x-\xi)$.

Problem: Second-order Equation (Damped harmonic oscillator)

$$d^2y/dt^2 + 2\ dy/dt + \omega_0^2 y = f(t)/m$$

$$d^2G(t, \tau)/dt^2 + 2\lambda dG(t, \tau)/dt + \omega_0^2 G(t, \tau) = \delta(t-\tau)/m$$

in the sense of distributions. Verify that G given by

$$G(t, \tau) = (1/\omega)\exp\{-\lambda(t-\tau)\}\sin\omega(t-\tau) \tag{4}$$

does solve the problem.

Problem: Beginning with the equation satisfied by the Green's function for the differential equation $Ly = f(x)$ with L defined by

$$L = (d/dx)[p(x)(d/dx)] - s(x) \text{ or } (d/dx)[p(x)(dG/dx)] - s(x)G = \delta(x-\xi)$$

show the magnitude of the jump discontinuity at $x = \xi$ is $1/p(\xi)$ by integrating the above equation between $\xi - \varepsilon$ and $\xi + \varepsilon$.

Problem: Show that the G in Example 5 satisfies the nonhomogeneous differential equation and boundary conditions by writing $y = \int_a^x G(x, \xi)f(\xi)\, d\xi + \int_x^b G(x, \xi)f(\xi)\, d\xi$ and doing the appropriate differentiations for substitution into the differential equation.

Any problem with nonhomogeneous boundary conditions can be changed into a problem with homogeneous conditions. Thus it is sufficient to consider problems with homogeneous conditions. The following example will clarify the technique:

Example 6. Consider the equation $d^2u/dx^2 + k^2u = -f(x)$ and two-point boundary conditions $u(0) = a$ and $u'(1) = b$. First, let's find the solution for homogeneous boundary conditions $u(0) = 0$ and $u'(1) = 0$. We will use the technique of example 3. The homogeneous equation

$$\frac{d^2G}{dx^2} + k^2G = 0$$

has solutions ϕ_1, ϕ_2. For $x < \xi$, $\phi_1 = \sin kx$ which satisfies $G(0, \xi) = 0$. For $x > \xi$, we must satisfy $G_x(1, \xi) = 0$ hence $\phi_2 = \cos k(1-x)$. We can either evaluate constants of proportionality by using the continuity and derivative conditions, or simply multiply ϕ_1 by the value of ϕ_2 evaluated at $x = \xi$ and ϕ_2 by the value of ϕ_1 evaluated at $x = \xi$. Then

$$G(x, \xi) = \sin kx \cos k(1-\xi) \qquad x < \xi$$

$$= \cos k(1-x) \sin k\xi \qquad x < \xi$$

If we take the difference of the derivatives at $x = \xi$ we get a jump of $-k \cos k$ instead of -1. Therefore we divide the above G by $k \cos k$ and have

$$G(x, \xi) = \frac{\sin kx \cos k(1-\xi)}{k \cos k} \qquad x < \xi$$

$$= \frac{\cos k(1-x) \sin k\xi}{k \cos k} \qquad x > \xi$$

and finally

$$G(x, \xi) = \frac{\sin kx \cos k(1-\xi)}{k \cos k} H(\xi-x) +$$

$$+ \frac{\cos k(1-x) \sin k\xi}{k \cos k} H(x-\xi)$$

for all x.

Before going on to multidimensional equations it may be valuable to summarize some ideas.

11. PROPERTIES OF OPERATOR L

We are discussing linear differential operators of the form $L = a_n(x)d^n/dx^n + a_{n-1}d^{n-1}/dx^{n-1} + \cdots + a_0(x)$ and equations of the form $Lu(x) = f(x)$. The coefficients $a_i(x)$ for $0 \leq i \leq n$ are assumed continuous on a closed interval I which may be [a, b] or conveniently [0, 1] on the x axis. This is indicated as $a_i(x) \in C^0(I)$ for $0 \leq i \leq n$ or, simply $a_i(x) \in C^0$ if it isn't necessary to be explicit about the interval I.

Similarly $C^n(I)$ for $n \geq 1$ represents the class of functions $f(x) \in C^0(I)$ which have the property that df/dx exists and belongs to $C^{n-1}(I)$. Thus the statement $f(x) \in C^n(I)$ implies that f(x) and its first n derivatives exist and are continuous on I.

12. ADJOINT OPERATORS

If u(x) and v(x) are of class C^k on I and x_0, x are points in I, then

$$\int_{x_0}^{x} v(\xi)u^{(k)}(\xi)\, d\xi$$

may be transformed into an integral in which the differentiation acts on v(x) instead of u(x) by repeated integrations by parts leading to

$$\sum_{\nu=0}^{k-1} (-1)^{\nu} v^{(\nu)}(\xi)u^{(k-1-\nu)}(\xi) \Big|_{x_0}^{x} + (-1)^{k} \int_{x_0}^{x} v^{(k)}(\xi)u(\xi)\, d\xi$$

Thus for an operator L

$$\int_{x_0}^{x} v(\xi)Lu(\xi)\, d\xi = \int_{x_0}^{x} u(\xi)L^{*}v(\xi)\, d\xi + \phi[u, v]\Big|_{x_0}^{x}$$

where L^* is the *adjoint operator* and $\phi[u, v]\Big|_{x_0}^{x}$ representing the sum in the equation above is called the *bilinear concomitant*. In the form

$$\int_{x_0}^{x} [vLu - uL^{*}v]\, d\xi = \phi[u, v]\Big|_{x_0}^{x}$$

this is *Green's formula*. Differentiating both sides yields *Lagrange's identity*

$$vLu - uL^{*}v = \frac{d}{dx}\phi[u, v]$$

On can show by the elementary method of variation of parameters that the nonhomogeneous equation Lu = f(x) for a second order differential operator that the solution can be given in the form

$$u(x) = c_1\phi_1(x) + c_2\phi_2(x) + \int_{x_0}^{x} G(x, \xi)f(\xi)\, d\xi$$

where ϕ_1, ϕ_2 are linearly independent solutions of the homogeneous equation. It helps understanding the Green's function concept to carry this out. Given the solutions ϕ_1, ϕ_2 of $Lu = 0$ where $L = a_2(x)d^2/dx^2 + a_1(x)d/dx + a_0(x)$, assume

$$u(x) = \alpha_1(x)\phi_1(x) + \alpha_2(x)\phi_2(x)$$

where $\alpha_1, \alpha_2 \in C^1(I)$. We are simply writing a linear combination of the homogeneous solutions but instead of constant coefficients α_1, α_2 we write $\alpha_1(x)$, $\alpha_2(x)$. Now

$$u'(x) = \alpha_1\phi_1' + \alpha_2\phi_2' + [\alpha_1'\phi_1 + \alpha_2'\phi_2]$$

We impose two conditions. The first of these is that

$$[\alpha_1'\phi_1 + \alpha_2'\phi_2] = 0$$

The second derivative is

$$u''(x) = \alpha_1\phi_1'' + \alpha_2\phi_2'' + [\alpha_1'\phi_1' + \alpha_2'\phi_2'] = f(x)$$

As a second condition, require that

$$[\alpha_1'\phi_1' + \alpha_2'\phi_2'] = f(x)/a_2(x)$$

If α_1, α_2 are chosen to satisfy these two conditions then,

$$\begin{aligned} Lu &= a_2(\alpha_1\phi_1'' + \alpha_2\phi_2'' + \frac{f}{a_2}) + a_1(\alpha_1\phi_1' + \alpha_2\phi_2') + \\ &\quad + a_0(\alpha_1\phi_1 + \alpha_2\phi_2) \\ &= \alpha_1 L\phi_1 + \alpha_2 L\phi_2 + f = f \end{aligned}$$

since $L\phi_1 = L\phi_2 = 0$. These two conditions imposed on the α's are a system of linear equations for α_1', α_2'. The solutions are

$$\alpha_1'(x) = -\phi_2(x)f(x)/a_2(x)W(x)$$

$$\alpha_2'(x) = \phi_1(x)f(x)/a_2(x)W(x)$$

where $W(x)$ is the Wronskian of $\phi_1(x)$ and $\phi_2(x)$ given by

$$W(x) = \begin{vmatrix} \phi_1(x) & \phi_2(x) \\ \phi_1'(x) & \phi_2'(x) \end{vmatrix}$$

Consequently

$$\alpha_1(x) = -\int_{x_0}^{x} \frac{\phi_2(\xi)f(\xi)}{a_2(\xi)W(\xi)}\, d\xi + c_1$$

$$\alpha_2(x) = \int_{x_0}^{x} \frac{\phi_1(\xi)f(\xi)}{a_2(\xi)W(\xi)}\, d\xi + c_2$$

where c_1, c_2 are constants and x, $x_0 \in I$. Now that we have α_1, α_2 we can substitute them into our assumed solution $u = \alpha_1\phi_1 + \alpha_2\phi_2$ to get

$$u(x)=c_1\phi_1(x)+c_2\phi_2(x)-\int_{x_0}^{x} \frac{[\phi_1(x)\phi_2(\xi)-\phi_2(x)\phi_1(\xi)]}{a_2(\xi)W(\xi)}\, f(\xi)\, d\xi$$

where $c_1\phi_1 + c_2\phi_2$ is the homogeneous solution and the integral kernel of the last term is the Green's function, i.e., identifying

$$G(x, \xi) = -\frac{1}{a_2(\xi)} \frac{\begin{vmatrix} \phi_1(x) & \phi_2(x) \\ \phi_1(\xi) & \phi_2(\xi) \end{vmatrix}}{\begin{vmatrix} \phi_1(\xi) & \phi_2(\xi) \\ \phi_1'(\xi) & \phi_2'(\xi) \end{vmatrix}}$$

we have

$$u(x) = c_1\phi_1(x) + c_2\phi_2(x) + \int_{x_0}^{x} G(x, \xi)f(\xi)\, d\xi$$

Thus realizing we get this form, then we can seek methods of finding the kernel G directly.

Example: Use the method of variation of parameters to find the particular integral for the equation $d^2u/dt^2 - u = f(t)$ and identify the Green's function.

13. n-th ORDER OPERATORS

If we consider the n-th order differential operator

$$L = \sum_{\nu=0}^{n} a_\nu(x) d^\nu/dx^\nu$$

where $a_i(x) \in C^0(I)$ for $0 \le i \le n$ and $a_n(x) > 0$ on I where I is a closed interval $[x_0, x_1]$ or $[a, b]$, or for convenience, $[0, 1]$ on the x axis, and where we have a fundamental set of solutions $\{\phi_j(x)\}$, $1 \le j \le n$ of the homogeneous equation $Lu = 0$. Then

$$G(x, \xi) = \frac{(-1)^{n-1}}{a_n(\xi)W(\xi)} \begin{vmatrix} \phi_1(x) & \phi_2(x) & \cdots & \phi_n(x) \\ \phi_1(\xi) & \phi_2(\xi) & \cdots & \phi_n(\xi) \\ \phi_1^1(\xi) & \phi_2^1(\xi) & \cdots & \phi_n^1(\xi) \\ \vdots & & & \\ \phi_1^{(n-2)}(\xi) & \phi_2^{(n-2)}(\xi) & \cdots & \phi_n^{(n-2)}(\xi) \end{vmatrix}$$

where W is the Wronskian of the $\phi_j(\xi)$.

If initial conditions are all zero, i.e., $u^{(k)}(x_0) = 0$ for $0 \le k \le n-1$, then $u = \int_{x_0}^{x} G(x, \xi)f(\xi)d\xi$ is a solution of $Lu = f(x)$.

Further reading regarding properties and theorems on this concept for those interested is suggested in books such as Linear Differential Equations by K.S. Miller (Norton, 1963) and Green's Functions by G.F. Roach (Van Nostrand, 1970) or the first chapter of Stochastic Systems by G. Adomian (Academic Press, 1983). Also an excellent Soviet reference is now available called Green's Functions and Transfer

Functions Handbook by A.G. Butkovsky (transl. by L.W. Langdon; Halsted Press, a division of Wiley). We recommend it strongly as a reference.

14. BOUNDARY CONDITIONS FOR THE STURM-LIOUVILLE EQUATION

An important form of the second-order differential equation is

$$Lu = p(x) \frac{d^2u}{dx^2} + p'(x) \frac{du}{dx} - q(x)u = -f(x)$$

defined on $[x_0, x_1]$ with p, p', q continuous functions on the interval $[x_0, x_1]$ with $p > 0$ and $f(x)$ a piecewise continuous function in $[x_0, x_1]$. Preferably we write

$$Lu = f(x)$$

with L defined by

$$L = -\frac{d}{dx} [p(x) \frac{d}{dx}] + q(x)$$

with boundary conditions $B_1(u) = a$, $B_2(u) = b$. Such two-point boundary conditions can be given as

$$B_i(u) = \sum_{j=1}^{n} a_{ij}\, u^{(j-1)}(x_0) + \sum_{j=1}^{n} b_{ij}\, u^{(j-1)}(x_1)$$

where the index i specifies the number of such conditions. Two-point boundary conditions imply that not all of the a_{ij} and not all of the b_{ij} are zero. For example,

$$B_i(u) = a_{10}u(0) + a_{11}u'(0) + b_{10}u(1) + b_{11}u'(1) = 0$$

$$B_2(u) = a_{20}u(0) + a_{21}u'(0) + b_{20}u(1) + b_{21}u'(0) = 0$$

or simply

$$B_1(u) = u(0) = 0$$

$$B_2(u) = u(1) = 0$$

are two-point boundary conditions.

15. GREEN'S FUNCTION FOR STURM-LIOUVILLE OPERATOR

Consider

$$L = -\frac{d}{dx}\left(p(x)\frac{d}{dx}\right) + q \qquad p \neq 0 \quad \text{on} \quad I(0, 1)$$

$$B_1(u) = B_2(u) = 0$$

where we assume B_1 involves values only at $x = 0$ and B_2 involves values only at $x = 1$. Then

$$LG(x, \xi) = \delta(x-\xi)$$

or

$$(pG_x)_x - qG = -\delta(x-\xi)$$

and (1)

$$B_1(G) = B_2(G) = 0$$

Suppose we solve $Lu = 0$ to get two linearly independent solutions v_1, v_2. Let $w_1(x)$ be a linear combination of v_1, v_2 satisfying $B_1(w_1) = 0$ and w_2 a linear combination of v_1, v_2 satisfying $B_2(w_2) = 0$. Try $G = w_1(x)$ for $x < \xi$ and $w_2(x)$ for $x > \xi$. We will thus satisfy (1) and the boundary conditions but not continuity at ξ or the jump condition of magnitude $-1/p(x)$. Multiply the solution for $x < \xi$ by the value of the solution for $x > \xi$ evaluated at ξ and vice versa. Thus a new trial Green's function is

$$G(x, \xi) = w_1(x)w_2(\xi) \qquad x < \xi$$

$$= w_2(x)w_1(\xi) \qquad x > \xi$$

Evaluate now the magnitude of the jump

$$\frac{dG}{dx} = \frac{d}{dx}[w_1(x)w_2(\xi)] = w_1'(x)w_2(\xi) \qquad x < \xi$$

$$\frac{dG}{dx} = \frac{d}{dx}[w_2(x)w_1(\xi)] = w_2'(x)w_1(\xi) \qquad x > \xi$$

The jump is the difference evaluated at ξ or

$$w_2'(\xi)w_1(\xi) - w_1'(\xi)w_2(\xi)$$

must be $-1/p(\xi)$ hence we must again correct the above G and write

$$G = \frac{p(\xi)[w_1(x)w_2(\xi)]}{w_2'(\xi)w_1(\xi) - w_1'(\xi)w_2(\xi)} \quad \text{for } x < \xi$$

$$= \frac{p(\xi)[w_2(x)w_1(\xi)]}{w_2'(\xi)w_1(\xi) - w_1'(\xi)w_2(\xi)} \quad \text{for } x > \xi$$

and finally

$$G(x, \xi) = \frac{p(\xi)}{[w_2'(\xi)w_1(\xi) - w_1'(\xi)w_2(\xi)]} \{w_1(x)w_2(\xi)H(\xi-x) +$$

$$+ w_2(x)w_1(\xi)H(x-\xi)\}$$

16. SOLUTION OF THE INHOMOGENEOUS EQUATION

We know that the Green's function satisfies

$$G'' + p(x)G' + q(x)G = -\delta(x-x') \tag{1}$$

for the inhomogeneous equation

$$y'' + p(x)y' + q(x)y = -f(x) \tag{2}$$

Multiplying (1) by $-y$ and (2) by G and adding,

$$G(y'' + py' + qy) - y(G'' + pG' + qG)$$

$$= -Gf + y\delta(x-x') \tag{3}$$

$$= (Gy'' - yG'') + p(Gy' - yG').$$

Since $Gy'' - yG'' = (d/dx)(Gy'-yG')$,

$$(d/dx)(Gy'-yG') + p(Gy'+yG') = -Gf+y\delta(x-x'). \tag{4}$$

The quantity $Gy'-yG'$ is the Wronskian W of G and y hence

$$\frac{dW}{dx} + p(x)W(G, y; x) = -G(x, x')f(x) + y(x)\delta(x-x') \qquad (5)$$

or

$$(d/dx)(W \exp \{\int_{x_0}^{x} p(\zeta)\, d\zeta\}) = \exp \{\int_{x_0}^{x} p(\zeta)\, d\zeta\}[\frac{dW}{dx} + p(x)W]$$

$$= -G(x, x') \exp \{\int_{x_0}^{x} p(\zeta)\, d\zeta\}f(x)+y(x) \exp \{\int_{x_0}^{x} p(\zeta)\, d\zeta\}\delta(x-x'). \qquad (6)$$

Integrating from x = a to x = b,

$$\int_a^b y(x) \exp \{\int_{x_0}^{x} p(\zeta)\, d\zeta\}\delta(x-x')\, dx = \int_a^b G(x, x')f(x) \exp \int_{x_0}^{x} p(\zeta) d\zeta dx +$$

$$+ w \exp \{\int_{x_0}^{x} p(\zeta)\, d\zeta\}\Big|_a^b . \qquad (7)$$

If x' lies inside [a, b] we have

$$y(x') = \int_a^b G(x, x')f(x) \exp \{-\int_{x_0}^{x'} p(\zeta)\, d\zeta + \int_{x_0}^{x} p(\zeta)\, d\zeta\}\, dx +$$

$$+ W(G, y) \exp \{-\int_{x_0}^{x'} p(\zeta)\, d\zeta + \int_{x_0}^{x} p(\zeta)\, d\zeta\}\Big|_{x=a}^{x=b} \qquad (8)$$

The last term can be written

$$W(G, y) \exp \{-\int_{x_0}^{x'} p(\zeta)\, d\zeta + \int_{x_0}^{x} p(\zeta)\, d\zeta\}\Big|_a^b$$

(equation continued on next page)

$$= [G(b, x')y'(b)-G'(b, x')y(b)] \exp\{-\int_{x_0}^{x'} p(\zeta)d\zeta + \int_{x_0}^{b} p(\zeta)\, d\zeta\} -$$

$$-[G(a, x')y'(a)-G'(a, x')y(a)] \exp\{-\int_{x_0}^{x} p(\zeta)\, d\zeta + \int_{x_0}^{a} p(\zeta)\, d\zeta\}. \tag{9}$$

Thus (8) represents $y(x')$, i.e., $y(x)$ at the interior point x' in $[a, b]$ in terms of boundary values of y and y' at a and b and in terms of the convolution integral we have seen before involving the Green's function $G(x, x')$ and the forcing function.

A natural question is the following. Suppose we are not given y and y' at a and at b. What if we are given $y(a)$, $y(b)$ or say $y'(a)$, $y'(b)$, or a linear combination $y' + \mu y$ at the endpoints a, b? The answer is to eliminate the terms that we don't have from our solution.

Example: Given $y(a) = \alpha$ and $y(b) = \beta$ we must choose c_1, c_2 in

$$G(x, x') = c_1G_1(x) + c_2G_2(x) + G(x, x')$$

so the solution of

$$G'' + p(x)G' + q(x)G = -\delta(x-x')$$

satisfies the homogeneous boundary conditions $G(a, x') = 0$ and $G(b, x') = 0$. Then only known quantities remain in (8) and (9).

Similarly if we know the boundary conditions $y'(a) = \alpha$, $y'(b) = \beta$, we choose G so it satisfies the homogeneous boundary conditions $G'(a, x') = 0$ and $G'(b, x') = 0$. Finally given $y(a) = \mu y'(a) = \alpha$ and $y(b) + \mu y'(b) = \beta$, with μ a constant, choose $G(x, x')$ satisfying homogeneous boundary conditions

$$\begin{aligned} G(a, x') + \mu G'(a, x') &= 0 \\ G(b, x') + \mu G'(b, x') &= 0 \end{aligned} \tag{10}$$

In all cases the Green's function satisfies a homogeneous boundary condition of the same form as the inhomogeneous condition satisfied by $y(x)$.

17. SOLVING NON-HOMOGENEOUS BOUNDARY CONDITIONS

Given the equation $Lu = f(x)$ with boundary conditions $B_1(u) = a$ and $B_2(u) = b$, we solve instead

$$Lu_1 = f(x)$$

$$B_1(u_1) = B_2(u_2) = 0$$

then find u_2 satisfying

$$Lu_2 = 0$$

$$B_1(u_2) = a$$

$$B_2(u_2) = b$$

Then $u = u_1 + u_2$ is the solution of the original equation.
If for example $L = d^2/dx^2$

$$u''(x) = f(x)$$

$$u(0) = a$$

$$u(1) = b$$

Let $u = u_1 + u_2$ and solve

$$u_1'' = f(x)$$

$$u_1(0) = u_1(1) = 0$$

and

$$u_2'' = 0$$

$$u_2(0) = a$$

$$u_2(1) = b$$

Exercise: Show solution is $u = \int_0^x (x-\xi)f(\xi)d\xi - x\int_0^1 (1-\xi)\, d\xi.$

Let $y_1(x)$, $y_2(x)$ be linearly independent solutions of the homogeneous Sturm-Liouville equation with $y_1(x)$ satisfying the boundary condition at $x = a$ and let $y_2(x)$ satisfying the boundary condition at $x = b$. Since $y_1(x)$ and $G_1(x, \xi)$ satisfy the same equation in the interval $[a, \xi]$, we have

$$G_1(x, \xi) = c_1 y_1(x)$$

where c_1 is a constant. Similarly since $y_2(x)$ and $G_2(x, \xi)$ satisfy the boundary condition at $x = b$, we have

$$G_2(x, \xi) = c_2 y_2(x)$$

Thus $LG_1 = 0$ for $x < \xi$ and $LG_2 = 0$ for $x > \xi$. The continuity condition at $x = \xi$ requires

$$c_2 y_2(\xi) - c_1 y_1(\xi) = 0$$

The discontinuity condition on the derivative of G requires that

$$c_2 y_2'(\xi) - c_1 y_1'(\xi) = -1/p(\xi)$$

These last two equations can now be solved for c_1, c_2. A unique solution exists provided the Wronskian $W(y_1, y_2)$ or

$$\begin{vmatrix} y_1(\xi) & y_2(\xi) \\ y_1'(\xi) & y_2'(\xi) \end{vmatrix} = y_1 y_2' - y_2 y_1' \neq 0$$

and this is the case since we assumed linearly independent solutions. Now since y_1 and y_2 satisfy $Ly = 0$

$$(py_1')' + qy_1 = 0$$

$$(py_2')' + qy_2 = 0$$

Multiply the first equation by y_2 and the second by y_1 and subtract to get

$$y_1(py_2')' - y_2(py_1')' = [p(y_1 y_2' - y_2 y_1']' = 0$$

so

$$p(y_1 y_2' + y_2 y_1') = pW(y_1, y_2) = a$$

where A is a constant (Abel's constant).

The discontinuity condition on the derivative of G requires that

$$c_2 y_2'(\xi) - c_1 y_1'(\xi) = -1/p(\xi)$$

The continuity of G at $x = \xi$ leads to

$$c_2 y_2(\xi) - c_1 y_1(\xi) = 0$$

Consequently

$$c_1 = -y_2(\xi)/A \quad \text{and} \quad c_2 = -y_1(\xi)/A$$

Now

$$G = c_1 y_1(x) = -y_2(\xi) y_1(x)/A \qquad x < \xi$$

$$G = c_2 y_2(x) = -y_1(\xi) y_2(x)/A \qquad x > \xi$$

where $A = pW(y_1, y_2)$. Thus

$$G(x, \xi) = \frac{y_1(x) y_2(\xi) H(\xi - x) + y_2(x) y_1(\xi) H(x - \xi)}{pW}$$

18. BOUNDARY CONDITIONS SPECIFIED ON FINITE INTERVAL [a, b]

For the Sturm-Liouville operator problem and boundary conditions on a finite interval [a, b] specified by

$$\alpha_1 y(a) + \alpha_2 y'(a) = 0 \quad \text{and} \quad \beta_1 y(b) + \beta_2 y'(b) = 0$$

where the α's are not both zero and the β's also are not both zero[1]. Let's consider the regions $a \leq x < \xi$ and $\xi < x \leq b$ separately. In the first (left) region, a solution of the homogeneous equation $Ly = 0$ will be denoted by y_1 which must of course satisfy the boundary condition at x=a or

$$\alpha_1 y_1(a) + \alpha_2 y_1'(a) = 0 \tag{1}$$

The Green's function G(x, ξ) also satisfies this condition hence

1) (Notes are at the end of the chapter, see page 235.)

$$\alpha_1 G(a, \xi) + \alpha_2 G(a, \xi) = 0 \tag{2}$$

Hence the Wronskian W(y, G) must be zero at x = a.

$$\begin{vmatrix} y & G \\ y' & G' \end{vmatrix} = 0 \tag{3}$$

or

$$y_1(a)G'(a, \xi) - y_1'(a)G(a, \xi) = 0 \tag{4}$$

Actually the Wronskian must be zero in the entire first region since W and G satisfy the same differential equation so they are not independent and $G(x, \xi) = c_1 y_1(x)$ for $0 \le x < \xi$ where c_1 is a constant.

Similarly if y_2 is a solution of the homogeneous equation and the condition at y = b then $G(x, \xi) = c_2 y_2(x)$ for $\xi < x \le b$. From our requirements of continuity of G and the discontinuity of G' at $x = \xi$,

$$\begin{aligned} c_1 y_1(\xi) - c_2 y_2(\xi) &= 0 \\ c_1 y_1'(\xi) - c_2 y_2'(\xi) &= -1/p(\xi) \end{aligned} \tag{5}$$

which we must solve for c_1 and c_2. As long as y_1 and y_2 are linearly independent solutions (so one is not a multiple of the other, i.e., if no solution other than zero of the homogeneous equation satisfies both boundary conditions at the same time) then we can find the c's. The solution is

$$c_1 = y_2(\xi)/p(\xi)W(\xi)$$

$$c_2 = y_1(\xi)/p(\xi)W(\xi)$$

where $W(\xi)$ or specifically $W(y_1, y_2; \xi)$ is the Wronskian of y_1 and y_2 at $x = \xi$ and $p(\xi)W(\xi)$ is simply a constant. Now

$$\begin{aligned} G(x, \xi) &= \frac{y_1(x)y_2(\xi)}{p(\xi)W(\xi)} \qquad a \le x < \xi \\ g(x, \xi) &= \frac{y_2(x)y_1(\xi)}{p(\xi)W(\xi)} \qquad \xi < x \le b \end{aligned} \tag{6}$$

Obviously G is symmetric and unique. A different solution for either y_1 or y_2 will be a multiple of the original one; the Wronskian has the same multiplying factor and G remains the same.

What if there is a solution for $Ly = 0$, say y_0, which satisfies both boundary conditions?

$$\begin{aligned} \alpha_1 y_0(a) + \alpha_2 y_0'(a) &= 0 \\ \beta_1 y_0(b) + \beta_2 y_0'(b) &= 0 \end{aligned} \tag{7}$$

then y_1 and y_2 are both multiples of y_0 and their Wronskian vanishes so the Green's function does not exist.

Problem: Show the above G satisfies the nonhomogeneous differential equation and boundary conditions by writing $y = \int_a^x G(x, \xi)f(\xi)\, d\xi + \int_x^b G(x, \xi)f(\xi)\, d\xi$ and doing the appropriate differentiations for substitution into the differential equation.

Example: For $y'' + k^2 y = 0$ with $k = 2$ and $y(0) = y'(1) = 0$ whose solutions are sin 2x, cos 2x. Thus

$$\begin{aligned} G(x, \xi) &= \sin 2x \quad \text{for } x < \xi \\ &= \cos 2(1-x) \quad \text{for } x > \xi. \end{aligned}$$

To make G oontinuous, multiply the first G by the value of the second, etc.,

$$G(x, \xi) = \cos 2\,(1-\xi) \sin 2x \quad \text{for } x < \xi$$

$$G(x, \xi) = \sin 2\xi \cos 2\,(1-x) \quad \text{for } x > \xi.$$

The jump in the derivative gives

$$2 \sin 2(1-\xi) \sin 2\xi - 2 \cos 2\xi \cos 2\,(1-\xi)$$

$$= -2 \cos 2(1-\xi+\xi) = -2 \cos 2$$

while the correct jump should equal -1. Consequently we divide the solution by $2 \cos 2$ to get

$$u_1' = 2 \cos 2x \cos 2(1-\xi)$$

$$u_2' = 2 \sin 2\xi \cos 2(1-x)$$

Example: Consider the equation

$$u''(x) + \alpha^2 u(x) = \beta(x)$$

with boundary conditions specified by $u(0) = u(\ell)$ and $u'(0) = u'(\ell)$. The solutions of the homogeneous equation $u'' + \alpha^2 u = 0$ are given by $\sin \alpha x$ and $\cos \alpha x$. Thus

$$u(x) = A_1 \sin \alpha x + A_2 \cos \alpha x \qquad (0 \le x \le \xi)$$

$$u(x) = B_1 \sin \alpha x + B_2 \cos \alpha x \qquad (\xi \le x \le \ell)$$

$$A_2 = B_1 \sin \alpha\ell + B_2 \cos \alpha\ell \qquad u(0) = u(\ell)$$

$$A_1 = B_1 \cos \alpha\ell - B_2 \sin \alpha\ell \qquad u'(0) = u'(\ell)$$

$u(x)$ is continuous at $x = \xi$ hence

$$(B_1-A_1) \sin \alpha\xi + (B_2-A_2) \cos \alpha\xi = 0.$$

The condition on $u'(x)$ at $x = \xi$ yields

$$\alpha(B_1-A_1) \cos \alpha\xi - \alpha(B_2-A_2) \sin \alpha\xi = 1$$

Four equations determine A_1, A_2, B_1, B_2

$$A_1 = \frac{\sin \alpha\,[\xi-(\ell/2)]}{2\alpha \sin \alpha\ell/2} \qquad A_2 = \frac{\cos \alpha\,[\xi-(\ell/2)]}{2\alpha \sin \alpha\ell/2}$$

$$B_1 = \frac{\sin \alpha\,[\xi+(\ell/2)]}{2\alpha \sin \alpha\ell/2} \qquad B_2 = \frac{\cos \alpha\,[\xi+(\ell/2)]}{2\alpha \sin \alpha\ell/2}$$

Therefore the Green's function is given by

$$G(x, \xi) = \frac{\cos \alpha\,(\xi-x-\ell/2)}{2\alpha \sin \alpha\ell/2} \qquad 0 \le x \le \xi$$

$$G(x, \xi) = \frac{\cos \alpha\,(\xi-x+\ell/2)}{2\alpha \sin \alpha\ell/2} \qquad \xi \le x \le \ell.$$

G is symmetric (since L is self-adjoint). If we interchange x and ξ, we see the symmetric property of G. For the region $x > \xi$ we have

$$G(\xi, x) = \frac{\cos \alpha\ (\xi - x + \ell/2)}{2\alpha \sin \alpha\ell/2} .$$

Let G_I denote the Green's function for the region $0 \leq x < \xi$, and G_{II} the Green's function for $\xi < x \leq \ell$. Now $G_I(\xi, x) = G_{II}(x, \xi)$ since

$$G_I(x, \xi) = \frac{\cos \alpha(\xi - x - \ell/2)}{2\alpha \sin \alpha\ell/2}$$

implies

$$= \frac{\cos \alpha(x - \xi + \ell/2)}{2\alpha \sin \alpha\ell/2}$$

$$= G_{II}(\xi, x) = \frac{\cos \alpha(\xi - x - \ell/2)}{2\alpha \sin \alpha\ell/2}$$

is an even function.

Example: Consider the equation

$$dy/dt + (1/a(t))y = x(t)/a(t).$$

G must satisfy

$$dG/dt + (1/a(t))G = \delta(t-\tau).$$

Therefore for $t > \tau$

$$G(t, \tau) = (1/a(\tau)) \exp \{\int_\tau^t du/a(u)\} .$$

If $a(t) = T$, a constante, then

$$G(t-\tau) = (1/T) \exp \{(-t-\tau)/T\}.$$

Example: Find the Green's function $G(x, \xi)$ satisfying the system:

$$d^2G/dx^2 - c^2G = -\delta(x-\xi) \qquad x, \xi \in [0, \ell]$$

$$G(0, \xi) = G(\ell, \xi) = 0 \qquad c = \text{constant}$$

We know G is continuous at $x = \xi$, the jump in the derivative $\partial G/\partial x$ across $x = \xi$ is -1, and G satisfies the homogeneous equation if $x \neq \xi$. Denoting G for $x < \xi$ by G_1 and for $x > \xi$ by G_2, it can be shown that

$$G_1 = a_2 \sinh cx$$

$$G_2 = a_2 \sinh c(\ell-x)$$

Since $G_1 = G_2$ at $x = \xi$,

$$a_1 \sinh c\xi = a_2 \sinh c(\ell-\xi)$$

and because of the derivative condition

$$-a_2c \cosh c\ (\ell-\xi) - a_1c \cosh c\xi = -1$$

From these equations, the constants a_1, a_2 are found to be

$$a_1 = \frac{\sinh c\ (\ell-\xi)}{c \sinh c}$$

$$a_2 = \frac{\sinh c\xi}{c \sinh c\ell}$$

Hence

$$G(x, \xi) = \frac{1}{c \sinh c\ell} \{\sinh c(\ell-\xi) \sinh cxh(\xi-x).$$
$$+ \sinh c\xi \sinh c(\ell-x)H(x-\xi)\}$$

19. SCALAR PRODUCTS

Let S be a set or space of real continuous functions which are Lebesgue integrable[2] on the interval [0, 1] (i.e., an L_2 space). Now for $0 \leq x \leq 1$, a function $u(x) \in S$ if $\int_0^1 u^2(x)\, dx$ is finite. The scalar product of two functions

$u(x)$, $v(x)$ contained in S is given by $\int_0^1 u(x)v(x)\,dx$ and denoted by the scalar product (u, v). In the (L_2) space of complex-valued functions, we define (u, v) by $\int_0^1 u^*(x)v(x)\,dx$.. Sometimes it is convenient to introduce a weight function $r(x)$ into the definition of the scalar product on S such that $(u, v) = \int_0^1 u(x)v(x)r(x)\,dx$ with S denoting all u such that $\int_0^1 u^2(x)r(x)\,dx$ is finite.

If we consider the space to be that of all continuous functions, then a differential operator cannot be applied to any $u \in S$ since not all are differentiable, and even if they were, the resulting function may not be in S. Consequently we take for S all real continuous functions $u(x)$ which have continuous first and second derivatives and such that $Lu \in S$ if L is a differential operator of the form $L = a(x)d^2/dx^2 + b(x)d/dx + c(x)$ where a, b, c are continuous functions on [0, 1].

The *adjoint operator* $L^\dagger$ can be defined by the property that $(v, Lu) = (L^\dagger v, u)$.

Suppose for example $L = d/dx$ then (v, Lu) means $\int_0^1 v \frac{du}{dx}\,dx = vu\Big|_0^1 - \int_0^1 u \frac{dv}{dx}\,dx$. If $vu\Big|_0^1$ is zero then

$$(v, Lu) = -\int_0^1 u \frac{dv}{dx}\,dx = (L^\dagger v, u), \text{ i.e. } L^\dagger = -d/dx.$$

The term which vanishes will be a linear homogeneous function of u and v and their derivatives. Suppose $Lu = -(pu')' + qu$ and the interval is [a, b]. The term which vanishes is easily seen to be $p(x)W(u, v; x)\Big|_a^b$ where $W(u, v; x) = uv' - vu'$, i.e., the Wronskian of u and v.

If $L = L^\dagger$ then $(v, Lu) = (L^\dagger v, u) = (Lv, u)$ hence

$$\int_a^b (vLu - uLv)\,dx = 0 \tag{1}$$

for all u(x), v(x) contained in the space of functions on which L acts.

If we write $Lu(x) = p_0(x)(d^2/dx^2)u(x)+p_1(x)(d/dx)u(x)+$ $+p_2(x)u(x)$ for real coefficients $p_i(x)$ on [a, b] where the first 2-i derivatives of $p_i(x)$ are continuous on [a, b], and $p_0(x)$ is non-vanishing on [a, b] and we define

$L^\dagger u = (d^2/dx^2)[p_0(x)u(x)]-(d/dx)[p_1(x)u(x)]+p_2(x)u(x)$

$= p_0(x)(d^2u/dx^2)+[2p_0'(x)-p_1(x)](du/dx)+[p_0''(x)-p_1'(x)+p_2(x)]u,$

we get $L^\dagger = L$ if and only if $p_0'(x) = (d/dx)p_0(x) = p_1(x)$. Then

$$L^\dagger u = p_0u'' + p_1u' + p_2u \;. \tag{2}$$

For our own purposes later, it is to be noted by looking at Lu and $L^\dagger u$, if we want to move the derivatives to the left it can be done using the adjoint. If $Lu = p_1(x)(d/dx)u(x)$, $L^\dagger u = -(d/dx)[p_1(x)u(x)]$.

If $Lu = p_0(x)(d^2/dx^2)u(x)$, $L^\dagger u = (d^2/dx^2)[p_0(x)u(x)]$.

Operators for real physical problems are always self-adjoint so $\int_a^b (vLu - uLv)dx = 0$ for all u, v in S and L can be put into self-adjoint form (see completeness and closure properties, Courant, pp. 277-280).

20. USE OF GREEN'S FUNCTION TO SOLVE A SECOND-ORDER STOCHASTIC DIFFERENTIAL EQUATION[3]

Consider the equation:

$$d^2y/dt^2 + a_1(t, \omega)dy/dt + a_0(t, \omega)y = x(t, \omega)$$

a second-order differential equation where a_0, a_1, and $x(t, \omega)$ are stochastic processes defined on a suitable probability space Ω and index space T. Because of the stochastic coefficients, the differential operator is a *stochastic operator* of the type considered by Adomian (1983). We let

$$a_0 = 1 + \alpha_0(t, \omega) \qquad \langle a_0\rangle = 1 \qquad \langle \alpha_0\rangle = 0$$

$$a_1 = \alpha_1(t, \omega) \qquad \langle a_1\rangle = 0 \qquad \langle \alpha_1\rangle = 0$$

Thus a_0 has a deterministic part and a_1 does not. We can write this equation as $Ly + Ry = x$ where $L = (d^2/dt^2) + 1$, $R = \alpha_1(d/dt) + \alpha_0$. With this L, $y = L^{-1}x - L^{-1}Ry$ which can be written in terms of the Green's function $\ell(t, \tau)$ for $L = (d^2/dt^2) + 1$. Thus supposing $y(0) = y'(0) = 0$, we have

$$y = \int_0^t \ell(t, \tau)x(\tau)\, d\tau - \int_0^t \ell(t, \tau)[\alpha_1(\tau, \omega)(d/d\tau)+\alpha_0(\tau, \omega)]y(\tau)\, d\tau$$

To evaluate $\ell(t, \tau)$, we find the solutions of the homogeneous equation $Ly = 0$. These are $\phi_0(t) = \cos t$ and $\phi_1(t) = \sin t$. Then the Green's function is given by

$$\ell(t, \tau) = -\frac{1}{W(\tau)} \begin{vmatrix} \cos t & \sin t \\ \cos \tau & \sin \tau \end{vmatrix}$$

Since the Wronskian $W = 1$,

$$\ell(t,\tau) = \cos t \sin \tau - \sin t \cos \tau = \sin(\tau - t)$$

for $t \geq \tau$ and zero otherwise. The equation is now completely solvable by the decomposition method (Adomian, 1983) using the above Green's function $\ell(t, \tau)$. (The first term in the second integral is handled easily by use of the adjoint operator $R^\dagger$ or equivalently integrating by parts.) The idea of course is to move the $d/d\tau$ so it doesn't act on the y which will be subjected to the decomposition to solve the equation by the Adomian method of approximation. We get

$$y = \int_0^t \ell(t, \tau)x(\tau)\, d\tau - \int_0^t (d/d\tau)[\ell(t, \tau)\alpha_1(\tau)]y(\tau)\, d\tau - \\ - \int_0^t \ell(t, \tau)\alpha_0(\tau)y(\tau)\, d\tau$$

Now writing $y = \sum_{i=0}^{\infty} y_i$ where we will of course determine an approximate solution $\phi_n = \sum_{i=0}^{n-1} y_i$ we get

$$y_0 = \int_0^t \ell(t, \tau)x(\tau)\, d\tau$$

$$y_1 = -\int_0^t \ell(t, \tau)\alpha_1(\tau)y_0(\tau)\, d\tau - \int_0^t \ell(t, \tau)\alpha_0(\tau)y_0(\tau)\, d\tau$$

$$y_2 = -\int_0^t \ell(t, \tau)\alpha_1(\tau)y_1(\tau)\, d\tau - \int_0^t \ell(t, \tau)\alpha_0(\tau)y_1(\tau)\, d\tau$$

etc.

so we determine ϕ_n to some appropriate n. This does require computing the Green's function for $L = d^2/dt^2 + 1$. We can do the same problem with a simpler L with considerable advantage in simplicity and computability of the above integrals. To do this we let $L = d^2/dt^2$ and $L_1 = 1$ with R unchanged. This yields a simpler inverse and therefore easier integrals. Now we write

$$y = y_0 - L^{-1}L_1 y - L^{-1}Ry$$

$$= y_0 - L^{-1}y - L^{-1}Ry$$

where $y_0 = y(0) + ty'(0) + L^{-1}x$ in general, and simply $L^{-1}x$ for zero initial conditions. It's again convenient to use the adjoint operator $R^{\dagger}$ as before. The operator L^{-1} obviously signifies a double integration. Hence

$$y_1 = -L^{-1}y_0 - L^{-1}Ry_0$$

$$y_2 = -L^{-1}y_1 - L^{-1}Ry_1$$

etc.

<u>Exercise</u>: Evaluate the Green's function for $L = (d^2/dt^2)+1$ and zero initial conditions. Determine the solution of the first part above and compare to the results for the simpler Green's function. You can let α_0 and α_1 be constants, then

functions of time, and finally stochastic processes. x also can first be deterministic and finally stochastic. Then $\langle y\rangle$ and $R_y = \langle y(t_1)y(t_2)\rangle$ can be determined.

In the second procedure, i.e., if $L = d^2/dt^2+1$ and $x = 0$, $R = 0$, and initial conditions are given by $y(0) = 1$ and $y'(0) = 1$, we have

$$[d^2/dt^2 + 1] = 0$$

and the solution

$$y_0 = 1 + t$$

$$y_1 = -L^{-1}y_0 = -\iint(1+t)\ dt = -\ (t + \frac{t^2}{2})\ dt$$

$$= -\frac{t^2}{2!} - \frac{t^3}{3!}$$

etc.

Then $y = 1 + t - t^2/2! - t^3/3! + \cdots$

The series $t - t^3/3! + \cdots = \sin t$. The series $1 - t^2/2! + \cdots = \cos t$, i.e. $y = \sin t + \cos t$.

If we have $a_0 = 1$ and $a_1 = 0$ so $L = d^2/dt^2 + 1$ but $x \neq 0$,

$$d^2y/dt^2 + y = x$$

by finding the Green's function for $d^2/dt^2 + 1$, we can verify

$$y = \int_0^t \sin\ (\tau-t)x(\tau)\ d\tau$$

Alternatively use $L = d^2/dt^2$ and the decomposition method. Let $x = 1$ in the solution as well as the one above to verify results. Then for $L = d^2/dt^2$

$$Ly = x-y = 1-y$$

$$y = L^{-1}(1) - L^{-1}(y)$$

$$y = (t^2/2!) - L^{-1}(y_0+y_1+\cdots)$$

$$= (t^2/2!)-(t^4/4!)+(t^6/6!) - \cdots$$

From the first approach

$$\sin\,(\tau-t) = (\tau-t) - \frac{(\tau-t)^3}{3!} + \cdots$$

$$\int \sin\,(\tau-t)d\tau = \frac{t^2}{2!} - \frac{t^4}{4!} + \cdots$$

the same result.

21. USE OF GREEN'S FUNCTION IN QUANTUM PHYSICS

Consider a linear (Hermitian) time-independent differential operator $L(\bar{r})$ and the differential equation

$$\frac{i}{c}\frac{\partial\phi}{\partial t} - L(\bar{r})\phi = 0$$

The Green's function must satisfy

$$[\frac{i}{c}\frac{\partial}{\partial t} - L(\bar{r})]G(\bar{r},\bar{r}',\ t,t') = \delta(\bar{r}-\bar{r}')\delta(t-t')$$

subject to homogeneous boundary conditions on the surface of the domain of $\bar{r},\bar{r}'$. If c is real and positive, the above is a Schrödinger equation. If c is imaginary, we have a diffusion equation. This is discussed by Economou (1979).

22. USE OF GREEN'S FUNCTIONS IN TRANSMISSION LINES

Consider the transmission line excited by a distributed current source.

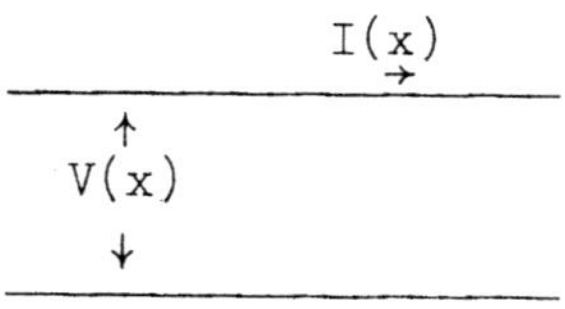

For a harmonically oscillating source, the voltage and the current on the line satisfy the equation:

$$\frac{dV(x)}{dx} = i\omega LI(x)$$

$$\frac{dI(x)}{dx} = i\omega CV(x) + K(x)$$

and eliminating I, we have

$$\frac{d^2V}{dx^2} + k^2V(x) = i\omega LK(x)$$

an inhomogeneous second-order scalar wave equation where $k = \omega\sqrt{LC}$ is the progagation constant, L is distributed inductance, C is distributed capacitance, and we can let $K(x) = (i/\omega L)\delta(x-\xi)$ to get a differential equation for the Green's function G. Thus

$$\frac{d^2G}{dx^2} + k^2G = -\delta(x-\xi)$$

Thus as in any network where we can determine the network response with an integration involving the impulse response (time-domain problem) we can now do the spatial equivalent with $G(x, \xi)$.

If the line is excited by a localized current source of amplitude $i\omega/L$ at $x = \xi$, and if we specify the boundary conditions, we obtain G and therefore the solution. The boundary conditions for G are the same as for V(x) and of course we get different G with different boundary conditions. We can consider the transmission line in the interval x_1, x_2 and imagine an impulse at the point $x = \xi$ in this interval. The boundary conditions depend on the terminations at the ends, i.e., the conditions on V(x) and dV(x)/dx determine G. If, for example, one end, say x = 0, is short-circuited, V = 0 at x = 0. If the line is open at $x = \infty$, we have $dV/dx - ikV = 0$ as the boundary condition. Such problems are discussed by Tai (1971).

23. TWO-POINT GREEN'S FUNCTIONS - GENERALIZATION TO n-POINT GREEN'S FUNCTIONS[4]

Consider the equation $dG/dt = \delta(t-\tau)$. Integrating with respect to t, we get $G = H(t-\tau)+\alpha(\tau)$. If we begin with $d^2G/dt^2 = \delta(t-\tau)$ we get $G = (t-\tau)H(t-\tau)+t\alpha_1(\tau)+\alpha_2(\tau)$. Similarly $d^3G/dt^3 = \delta(t-\tau)$ leads to $G = (t-\tau)^2/2!+(t^2/2)\alpha_1(\tau)+$ $+\alpha_3(\tau)$. Generalizing we write

$$\frac{d^n G}{dt^n} = \delta(t-\tau)$$

$$G = \frac{(t-\tau)^{n-1}}{(n-1)!} + \sum_{\nu=1}^{n} \alpha_\nu(\nu)\frac{t^{n-\nu}}{(n-\nu)!}, \quad n \geq 1 \text{ and } t > \tau$$

$$G = (t-\tau)^{n-1}/(n-1)! + \sum_{\nu=1}^{n} [t^{n-\nu}/(n-\nu)!]\alpha_\nu(\tau)$$

Thus the Green's function is determined for the nth order differential operator. This result will be very useful in our solutions of differential and partial differential equations by the decomposition method.

Let's consider d^2G/dt^2 and the result of two integrations:

$$G = (t-\tau)H(t-\tau) + t\alpha(\tau) + \beta(\tau)$$

Choose $u(0) = u(1) = 0$ as the boundary condition on the solution u of the equation $Lu = f(x)$ for which we are finding the Green's function. Then the Green's function satisfies $G(0,\tau) = G(1, \tau) = 0$. Thus

$$G(0, \tau) = (t-\tau)H(t-\tau) + t\alpha(\tau) + \beta(\tau) = 0$$

$$G(1, \tau) = (t-\tau)H(t-\tau) + t\alpha(\tau) + \beta(\tau) = 0$$

The condition on $G(0, \tau)$ requires that $\beta(\tau) = 0$. The condition on $G(1, \tau)$ requires that $(1-\tau)H(1-\tau) + \alpha(\tau) = 0$ or $\alpha(\tau) = (\tau-1)H(1-\tau)$ or $-(1-\tau)$ for $t \leq 1$. Thus

$$G(t, \tau) = (t-\tau)H(t-\tau)-t(1-\tau)$$

For d^3G/dt^3, we found three arbitrary functions so if we are given two conditions $u(t_1) = \beta_1$ and $u(t_2) = \beta_2$, the problem is undetermined. If we consider the nth order case,

$$d^nG/dt^n = \delta(t-\tau)$$

we found

$$G(t, \tau) = \frac{(t-\tau)^{n-1}}{(n-1)!} H(t-\tau) + \sum_{\nu=1}^{n} \alpha_\nu(\tau)\frac{t^{n-\nu}}{(m-\nu)!} \tag{1}$$

Suppose the boundary conditions are general two-point conditions rather than the simple conditions above. Thus,

$$\sum_{\nu=0}^{n-1} h_\nu \frac{d^\nu}{dt^\nu} u(t_1) = \beta_1$$

$$\sum_{\nu=0}^{n-1} k_\nu \frac{d^\nu}{dt^\nu} u(t_2) = \beta_2$$

The conditions on G are of course the same consequently

$$\sum_{\nu=0}^{n-1} h_\nu \frac{d^\nu}{dt^\nu} G(t_1, \tau) = \beta_1$$

$$\sum_{\nu=0}^{n-1} k_\nu \frac{d^\nu}{dt^\nu} G(t_2, \tau) = \beta_2$$

These are insufficient to determine the α_ν in (1) and therefore the complete Green's function to solve

$$Lu = \frac{d^n u}{dt^n} = f(t)$$

If we write boundary conditions

$$\sum_{i=0}^{n-1} k_{ij} \frac{d^i}{dt^i} u(t_j) = \beta_j$$

we have now n equations to determine the n α_ν's in (1) and the Green's function is completely determined from (1) and

$$\sum_{i=0}^{n-1} k_{ij} \frac{d^i}{dt^i} G(t_j, \tau) = \beta_j$$

Thus

$$\sum_{i=0}^{n-1} k_{ij} \left\{ \frac{(t-\tau)^{n-1-i}}{(n-1-i)!} H(t_j-\tau) + \sum_{\nu=1}^{n-i} \alpha_\nu(\tau) \frac{t^{n-\nu-i}}{(n-\nu-i)!} \right\} = \beta_j$$

$1 \le j \le n$

which gives us n equations in n unknowns to determine the α's. The k_{ij}, t_j, β_j are known from the specified boundary conditions. If the β_j are given as a column vector of n components and α_j as a similar column vector, the boundary condition is equivalent to an $n \times n$ matrix Γ multiplying the n vector α to give the n vector β', i.e., $\Gamma\alpha = \beta'$. Then $\alpha = \Gamma^{-1}\beta'$.

It is interesting to know that the inverse of a matrix can also be determined by Adomian's decomposition method. This is an approximation in n terms which uses as the first term of the approximate inverse a very simple invertible matrix. It will be discussed more appropriately in another publication. We mention it here as a matter of interest.

24. EVALUATION OF ARBITRARY FUNCTIONS FOR NONHOMOGENEOUS BOUNDARY CONDITIONS BY MATRIX EQUATIONS

For $Lu = g(x)$ with $L = d^2/dx^2$, we have

$$\begin{aligned} d^2G/dx^2 &= \delta(x-\xi) \\ G &= (x-\xi)H(x-\xi)+x\alpha_1(\xi)+\alpha_2(\xi) \end{aligned} \tag{1}$$

For the case of *homogeneous* boundary conditions $u(0) = u(1) = 0$ we have $G(0, \xi) = \alpha_2(\xi) = 0$ and $G(1, \xi) = 0 = (1-\xi)H(1-\xi)+\alpha_1(\xi)$ so that $\alpha_1(\xi) = -(1-\xi)H(1-\xi)$ and finally

$$G(x, \xi) = (x-\xi)H(x-\xi)-x(1-\xi)H(1-\xi)$$

for $x \in [0, 1]$ and $\xi \leq 1$. For *nonhomogeneous* boundary conditions, $u(b_1)= \beta_1$ and $u(b_2)= \beta_2$. Now using (1)

$$G(b_1, \xi) = (b_1-\xi)H(b_1-\xi)+b_1\alpha_1(\xi)+\alpha_2(\xi) = \beta_1$$

$$G(b_2, \xi) = (b_2-\xi)H(b_2-\xi)+b_2\alpha_1(\xi)+\alpha_2(\xi) = \beta_2$$

or

$$\begin{vmatrix} b_1 & 1 \\ b_2 & 1 \end{vmatrix} \cdot \begin{vmatrix} \alpha_1(\xi) \\ \alpha_2(\xi) \end{vmatrix} = \begin{vmatrix} \beta_1 \\ \beta_2 \end{vmatrix} - \begin{vmatrix} (b_1-\xi)H(b_1-\xi) \\ (b_2-\xi)H(b_2-\xi) \end{vmatrix}$$

from which α_1, α_2 are determined. The procedure will be generalized in a following section for partial differential equations.

25. MIXED BOUNDARY CONDITIONS

Considering the same equation as in the preceding section but imposing conditions on G given by

$$k_{11}\frac{dG(b_1, \xi)}{dx} + k_{12}G(b_1, \xi) = \beta_1$$

$$k_{21}\frac{dG(b_2, \xi)}{dx} + k_{22}G(b_2, \xi) = \beta_2$$

$$G(x, \xi) = (x-\xi)H(x-\xi) + x\alpha_1(\xi) + \alpha_2(\xi)$$

$$\frac{dG(x, \xi)}{dx} = H(x-\xi) + \alpha_1(\xi)$$

Consequently

$$k_{11}[H(b_1-\xi)+\alpha_1(\xi)]+k_{12}[(b_1-\xi)H(b_1-\xi)+b_1\alpha_1(\xi)+\alpha_2(\xi)] = \beta_1$$

$$k_{21}[H(b_2-\xi)+\alpha_1(\xi)]+k_{22}[(b_2-\xi)H(b_2-\xi)+b_2\alpha_1(\xi)+\alpha_2(\xi)] = \beta_2$$

$$\begin{vmatrix} k_{11} + k_{12}b_1 \\ k_{21} + k_{22}b_2 \end{vmatrix} \begin{vmatrix} \alpha_1(\xi) \\ \alpha_2(\xi) \end{vmatrix}$$

$$= \begin{vmatrix} \beta_1 \\ \beta_2 \end{vmatrix} - \begin{vmatrix} k_{11}H(b_1-\xi)+k_{12}(b_1-\xi)H(b_1-\xi) \\ k_{21}H(b_2-\xi)+k_{22}(b_2-\xi)H(b_2-\xi) \end{vmatrix}$$

which determines the α_1, α_2.

26. SOME GENERAL PROPERTIES

1. Nonnegativity of Green's Functions and Solutions

Various methods exist (Bellman, 1957a and b) of establishing non-negativity of the Green's function associated with ordinary differential equations of the form

$$u'' + q(x)u = f(x) \tag{1}$$

where $u(0) = u(1) = 0$. We present here another method which has certain merits. It indicates how useful it may be to study the behavior of the solution of $Lu = v$, where L is a linear operator by means of the limiting behavior of the solution of

$$\frac{\partial u}{\partial t} = Lu - v \tag{2}$$

as $t \to \infty$. This method has been used by Arrow and Hearon to study the inverse of input-output matrices (Bellman, 1960).

Consider the partial differential equaiton of parabolic type,

$$u_t + q(x)u - f(x) = u_{xx} \tag{3}$$

with the initial condition $u(x, 0) = h(x)$, with $h(x) \geq 0$, $0 \leq x \leq 1$, and $u(0, t) = u(1, t) = 0$. If we suppose that $q(x)$ is uniformly bounded, $0 \leq x \leq 1$, it is easy to show, under the hypotheses that $h(x)$, $f(x) \geq 0$, that $u(x, t) \geq 0$ for $t \geq 0$. If $q(x) \geq 0$, we use the finite difference approximation

$$u(x, t+\Delta^2) = \frac{u(x+\Delta, t)+u(x-\Delta, t)}{2} + q(x)u(x, t)\Delta + f(x)\Delta, \tag{4}$$

$t = 0, \Delta^2, \cdots$, which establishes inductively that $u(x, t) \geq 0$. As $\Delta \to 0$, the solution of the finite difference equation converges to that of the partial differential equation, thus establishing the required nonnegativity of the solution.

If $q(x)$ is not nonnegative, but bounded from below by a constant so that $M + q(x) \geq 0$, we write $u = e^{-Mt}w(x)$. Substituting, we obtain the equation

$$w_t = w_{xx} + (q(x) + M)w - f(x)e^{-Mt}, \tag{5}$$

which we can treat as in (6) to establish nonnegativity.

Returning to (3), we let $T \to \infty$. If all characteristic values of the equation

$$u_{xx} + q(x)u = \lambda u \tag{6}$$

with $u(0) = u(1) = 0$ are negative, the solution of (3) converges as $t \to \infty$ to the solution of (1) establishing thereby the nonnegativity of the solution of (1).

There is no difficulty in extending the proof to cover ordinary differential equations with more general boundary conditions and multidimensional partial differential equations of parabolic type.

2. Variation-Diminishing Properties of Green's Functions

The study of the properties of solutions of the boundary value problem

$$u'' + q(x)u = f(x). \tag{7}$$

$u(0) = u(1) = 0$, as the forcing term $f(x)$ ranges over all elements in a function class is equivalent to the study of properties of the associated Green's function $K(x, y)$. This function enables us to write the solution of (7) in the form

$$u(x) = \int_0^1 K(x, y)f(y)\, dy. \tag{8}$$

The question of determining the nonnegativity properties of $K(x, y)$ and more generally, the variation-diminishing properties, has been studied by a number of authors: Kellogg, Gantmakher and Krein (1960); Schoenberg (1953); Aronszajn and Smith (1956); Karlin and MacGregor (1958); and others. In a brief note (Bellman, 1957), it was shown that the nonnegativity of $K(x, y)$ could be deduced very simply from a consideration of the related quadratic form

$$J(u) = \int_0^1 [u'^2 + q(x)u^2 + 2f(x)u]\, dx \tag{9}$$

In this section, let us show that the variation-diminishing property of the Green's function follows in the same simple fashion from the problem of the determination of the absolute minimum of $J(u)$.

Variation Diminishing Property. The result we wish to establish is

THEOREM. Let q(x) satisfy the condition

$$q(x) \le \pi^2 - d, \quad d > 0 \tag{10}$$

where π^2 appears as the smallest characteristic value of the Sturm-Liouville problem

$$u'' + \lambda u = 0 \quad u(0) = u(1) = 0. \tag{11}$$

Then, if q(x) is as a continuous function with N changes of sign, the solution of (10) has at most N changes of sign.
Proof of Theorem. The Euler equation obtained from the first variation of J(u) over all functions which vanish at x = 0 and x = 1 is (7). For the sake of simplicity, take the simplest case where f(x) has one change of sign, and the solution u(x) is assumed to have two changes of sign, as indicated below.

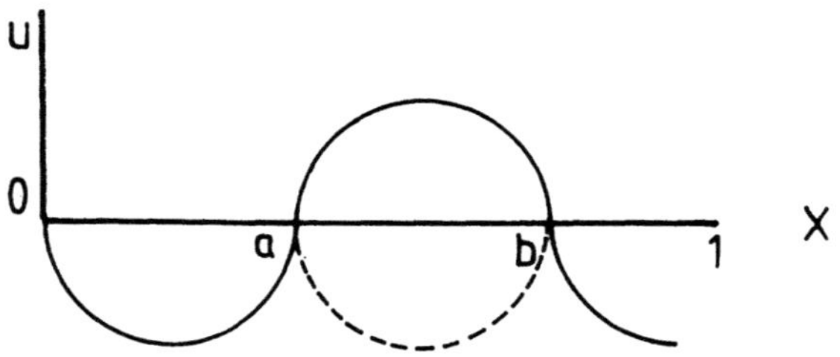

Were this case to hold, we could obtain a smaller value for J(u) by replacing u(x) by a function which has the same values in the interval [0, b], and is equal to the negative of u(x) in [b, 1]. This change leaves the quadratic terms unchanged, and decreases the contribution of the term 2f(x)u. This contradicts the fact that the solution of the Euler equation in this case yields the absolute minimum.

It is easy to see that the same argument handles the general case where any number of changes of sign is allowed. We shall discuss the multidimensional case elsewhere.

NOTES

1. Thus Dirichlet and Neumann conditions are included.
2. For our purposes, a Lebesgue integral is no different than the ordinary Riemann integral. The interested reader can refer to Titchmarsh, Theory of Functions (Clarendon Press, Chapter XI). The concept of the Lebesgue integral is quite important in analysis. It is an extension of the Riemann integral. We note that every Riemann integrable function is also Lebesgue integrable to the same value. There esists some functions which are not Riemann integrable but are Lebesgue integrable. The converse does not arise in physical problems.
3. See Stochastic Systems, G. Adomian (Academic Press, 1983) p. 142.
4. Also "two-sided" Green's functions. See Miller (1963).

BIBLIOGRAPHY

Alexsandrov, A.D., Chapter XIX in Mathematics, Its Content, Methods, and Meaning, Vol. III, M.I.T. Press, 1963.

Arfken, G., Mathematical Methods for Physicists, Academic Press, 1970.

Aronszajn, N. and Smith, K.T., A Characterization of Positive Reproducing Kernels. Application to Green's Functions Studies in Eigenvalue Problems, Technical Report 15, University of Kansas, 1956.

Beckenbach, E.F. and Bellman, R., Inequalities, Ergebnisse der Math., Springer, Berlin, 1961.

Bellman, R., 'On the Nonnegativity of Green's Functions', Boll U.M.I. 12 (1957) 411-413.

Bellman, R., 'On the Nonnegativity of the Heat Equation', Boll U.M.I. 12 (1957) 520-523.

Bellman, R., An Introduction to Matrix Analysis, McGraw-Hill, New York, 1960.

Birkhoff, G. and Rota, G.C., Ordinary Differential Equations, Blaisdell Co., 1969.

Butkov, E., Mathematical Physics, Addison-Wesley, 1968.

Courant-Hilbert, Methods of Mathematical Physics, Trans., New York Interscience Publishers, Vol. 1 & 2, 1953-.

Duff, G.F. and Naylor, D., Differential Equations of Applied Mathematics, John Wiley, 1966.

Economou, E.N., Green's Functions in Quantum Physics, Springer-Verlag, 1979.

Friedman, B., Principles of Applied Mathematics, Wiley, 1956.

Gantmakher, F.R. and Krein, M.G., Oszillationsmatrizen, Akademie-Verlag, Berlin, 1960.
Gelfand, I.M. and Vilenkin, N.Y. Generalized Functions, 4 vols., Academic Press, 1964.
Greenberg, M.D., Application of Green's Functions in Science and Engineering, Prentice Hall, 1971.
Karlin, S. and MacGregor, J., Coincidence Probabilities, Stanford University Technical Report 8, 1958.
Kraut, E.A., Fundamentals of Mathematical Physics, McGraw Hill, 1967.
Miller, K.S., Linear Differential Equations, Norton, 1963.
Morse and Feshbach, Methods of Theoretical Physics, McGraw Hill, 1953.
Pugachev, V.S., Theory of Random Functions, Addison-Wesley, 1965.
Roach, G.F., Green's Functions, Cambridge, 1982.
Schoenberg, I.J., 'On Smoothing Operations and Their Generating Functions', Bull. Amer. Math. Soc. 59 (1953) 199-230.
Tai, C.T., Dyadic Green's Functions in Electromagnetic Theory Intext, 1971.
Titchmarsh, E.C., Theory of Functions, Clarendon Press (2nd ed.), 1939.
Wylie, C.R., Advanced Engineering Mathematics, McGraw Hill, 1960.
Zemanian, A., Distribution Theory, McGraw Hill, 1965.

Chapter XVI

APPROXIMATE CALCULATION OF GREEN'S FUNCTIONS

The basic idea here is that a Green's function which may be difficult to determine in particular cases can be determined in an easily computable series by decomposition of a differential operator L, which is sufficiently difficult to merit approximation, into an operator L_1 whose inverse is known, or found with little effort, and an operator L_2 with no smallness restrictions, whose contribution to the total inverse L^{-1} can be found in series form.

Consider therefore a differential equation $Ly = x(t)$ where L is a linear deterministic ordinary differential operator of the form $L = \sum_{\nu=0}^{n} a_\nu(t)d^\nu/dt^\nu$ where a_n is non-vanishing on the interval of interest.

Decompose L into $L_1 + L_2$ where L_1 is sufficiently simple that determination of its Green's function is trivial. Then if L_2 is zero, we have simply $y(t) = \int_0^t \ell(t, \tau)x(\tau)\, d\tau$ where $\ell(t, \tau)$ is the Green's function for the L_1 operator. If L is a second-order differential operator we may have $L_1 = d^2/dt^2$ and L_2 will be the remaining terms of L say $\alpha(t)d/dt + \beta(t)$. More generally, $L = \sum_{\nu=0}^{n} a_\nu(t)d^\nu/dt^\nu$ and we might take $L_1 = d^n/dt^n$ and $L_2 = \sum_{\nu=0}^{n-1} a_\nu(t)d^\nu/dt^\nu$. We have

$$Ly = (L_1+L_2)y = x(t) \tag{1}$$

$$L_1y = x(t)-L_2y$$

$$y = L_1^{-1}x-L_1^{-1}L_2y \tag{2}$$

Now assume a decomposition $y = \sum_{i=0}^{\infty} y_i$ identifying y_0 as $L_1^{-1}x$ and write

$$y = L_1^{-1}x - L_1^{-1}L_2(y_0 + y_1 + \cdots) \tag{3}$$

from which we can determine the y_i - each being determinable in terms of the preceding y_{i-1}. Thus

$$y = L_1^{-1}x - L_1^{-1}L_2y_0 - L_1^{-1}L_2y_1 - \cdots$$

or

$$y = L_1^{-1}x - L_1^{-1}L_2L_1^{-1}x + L_1^{-1}L_2L_1^{-1}L_2L_1^{-1}x - \cdots \tag{4}$$

We have

$$y = \sum_{i=0}^{\infty} (-1)^i(L_1^{-1}L_2)^iL_1^{-1}x \tag{5}$$

thus

$$\begin{aligned} y_0 &= L_1^{-1}x \\ y_1 &= -L_1^{-1}L_2L_1^{-1}x \\ y_2 &= L_1^{-1}L_2L_1^{-1}L_2L_1^{-1}x \\ &\vdots \\ y_i &= (-1)^i(L_1^{-1}L_2)^iL_1^{-1}x\ . \end{aligned} \tag{6}$$

Hence the inverse of the differential operator L is given by

$$L^{-1} = \sum_{i=0}^{\infty} (-1)^i(L_1^{-1}L_2)^iL_1^{-1} \tag{7}$$

Equation (2) written out explicitly in terms of the Green's function for L_1^{-1} is

$$y = \int_0^t \ell(t, \tau)x(\tau)\ d\tau - \int_0^t \ell(t, \tau)L_2[y(\tau)]d\tau \tag{8}$$

$$= \int_0^t \ell(t, \tau)x(\tau)\ d\tau - \int_0^t L_2^{\dagger}[\ell(t, \tau)]\ y(\tau)d\tau \tag{9}$$

$$= \int_0^t \ell(t, \tau)x(\tau)\, d\tau - \int_0^t \sum_{i=0}^{\infty} (-1)^i \frac{d^i}{d\tau^i} [\ell(t, \tau)]\, d\tau \quad (10)$$

If $L_1 = a_n d^n/dt^n$ and $L_2 = \sum_{i=0}^{n-1} a_\nu(t) d^i/dt^i$

$$y(t) = \int_0^t \ell(t, \tau)x(\tau)\, d\tau - \int_0^t \sum_{i=0}^{n-1} (-1)^i \frac{d^i}{d\tau^i}[\ell(t, \tau)a_i(\tau)]y(\tau)\, d\tau$$

or

$$y(t) = \int_0^t \ell(t, \tau)x(\tau)\, d\tau - \int_0^t k(t, \tau)y(\tau)\, d\tau \quad (11)$$

where $k(t, \tau) = \sum_{i=0}^{n-1} (-1)^i \frac{d^i}{d\tau^i}[\ell(t, \tau)a_i(\tau)]$.

Let's choose i=0 i.e., we work with the case $L = L_1 + \alpha(t)$, i.e., $L_2 = \alpha(t)$. Now

$$y(t) = L_1^{-1}x - L_1^{-1}\alpha(t)L_1^{-1}x - L_1^{-1}\alpha L_1^{-1}\alpha L_1^{-1}x - \cdots \quad (12)$$

i.e.,

$$y_0 = \int_0^t \ell(t, \tau)x(\tau)\, d\tau$$

$$y_1 = -\int_0^t \ell(t, \tau)a(\tau) \int_0^\tau \ell(\tau, \gamma)x(\gamma)\, d\gamma\, d\tau$$

$$y_2 = \int_0^t \int_0^\tau \int_0^\gamma \ell(t, \tau)\ell(\tau, \gamma)\ell(\gamma, \sigma)a(\tau)a(\gamma)x(\sigma)\, d\gamma\, d\tau\, d\sigma$$

etc. Equivalently,

$$y(t) = \int_0^t \ell(t, \tau)x(\tau)\, d\tau - \int_0^t k(t, \tau)y(\tau)\, d\tau \quad (13)$$

where $k(t, \tau) = \ell(t, \tau)a(\tau)$. Hence

$$y(t) = \int_0^t \ell(t, \tau)x(\tau)\, d\tau - \int_0^t k(t, \tau)y_0(\tau)\, d\tau + $$
$$+ \int_0^t k(t, \tau)y_1(\tau)\, d\tau + \cdots$$

If we let $F(t) = L_1^{-1}x = \int_0^t \ell(t, \tau)x(\tau)\, d\tau$ we can write

$$y(t) = F(t) - \int_0^t k(t, \tau)F(\tau)\, d\tau + \int_0^t \int_0^\tau k(t, \tau)k(\tau, \gamma)F(\gamma)\, d\gamma\, d\tau - $$
$$- \int_0^t \int_0^\tau \int_0^\gamma k(t, \tau)k(\tau, \gamma)k(\gamma, \sigma)F(\sigma)\, d\sigma\, d\gamma\, d\tau + \cdots \quad . \tag{14}$$

If L_1 has constant coefficients, $\ell(t, \tau) = \ell(t-\tau)$. To make this clear, let us now consider the simple example $L = L_1+\alpha$ with α constant and $L_1^{-1} = \int dt$ and the Green's function $\ell = 1$, and proceed with our stated objective of determining the Green's function $G(t, \tau)$ for L. G satisfies $LG(t, \tau) = \delta(t-\tau)$ or

$$(L_1+L_2)G = \delta(t-\tau). \tag{15}$$

Thus G can be found from the preceding equations by replacing x by the δ function. For the last example $L = d/dt + \alpha$

$$(d/dt + \alpha)G(t, \tau) = \delta(t-\tau).$$

If we write $G = G_0+G_1+\cdots$ we have immediately (using $\ell = 1$, $L_2=\alpha$, $y=G$, $x=\delta(t, \tau)$)

$$G_0 = \int_{\tau}^{t} \delta(t - \tau)\, d\tau = 1 \qquad (t > \tau)$$

Remembering $k(t-\tau) = \ell(t-\tau)\alpha = \alpha$

$$G_1 = -\alpha\int_{\tau}^{t} d\tau = -\alpha(t-\tau)$$

$$G_2 = \int_0^t d\tau \int_0^{\tau} d\gamma\ \alpha^2 = \alpha^2(t-\tau)/2 \tag{16}$$

$$\vdots$$

Consequently

$$G = 1-\alpha(t-\tau)+\alpha^2(t-\tau)^2/2 + \cdots \tag{17}$$

an approximation to

$$G = e^{-\alpha(t-\tau)} \tag{18}$$

Physically this equation could model a particle of mass m moving as a result of a force f(t) in a resisting medium

$$m\ dv/dt + Rv = f(t)$$

or

$$(L+\alpha)v = f(t)/m$$

where $L = d/dt$ and $\alpha = R/m$. We have

$$(L+\alpha)G(t,\ \tau) = \delta(t-\tau)/m.$$

Now

$$G_0 = L_1^{-1}\,\frac{\delta(t-\tau)}{m} = 1/m, \qquad G_1 = -(\alpha/m)\int_0^t\int_0^{\tau} \delta(\gamma-\tau)\, d\gamma\, d\tau$$

$$= -(\alpha/m)(t-\tau),\ \cdots$$

Thus

$$G = 1/m \,[1 - \frac{R}{m}(t-\tau) + \cdots]$$

$$\simeq \frac{1}{m}\exp\{-(R/m)(t-\tau)\} \qquad t > \tau$$

Thus we use L_1^{-1} as a first approximation and find the total response function G as a series in which L_2 *need not be a perturbation* on L_1.

Exercise: $L_1 = d^2/dt^2$ and $L_2 = \alpha d/dt + \beta(t)$.

$$y = L_1^{-1}x - L_1^{-1}L_2L_1^{-1}x + L_1^{-1}L_2L_1^{-1}L_2L_1^{-1}x - \cdots$$

$G(t, \tau)$ satisfies

$$[d^2/dt^2 + \alpha d/dt + \beta]\; G(t, \tau) = \delta(t-\tau).$$

The Green's function $\ell(t, \tau)$ for $L_1 = d^2/dt^2$ with boundary conditions given as $G(0, \tau) = G'(0, \tau) = 0$ is $\ell(t, \tau)$. Find $G(t, \tau)$.

Problem: Let $L_1 = d^2/dt^2$ and $L_2 = \alpha d/dt + \beta(t)$. Then

$$y = L_1^{-1}x - L_1^{-1}L_2L_1^{-1}x + L_1^{-1}L_2L_1^{-1}L_2L_1^{-1}x - \cdots$$

and $G(t, \tau)$ satisfies

$$[d^2/dt^2 + \alpha d/dt + \beta]\; G(t, \tau) = \delta(t-\tau).$$

The Green's function $\ell(t, \tau)$ for $d^2G/dt^2 = \delta(t-\tau)$ with boundary conditions given as $G(0, \tau) = G'(0, \tau) = 0$ is $\ell(t, \tau)$. Find $G(t, \tau)$.

Chapter XVII

GREEN'S FUNCTIONS FOR PARTIAL DIFFERENTIAL EQUATIONS

1. INTRODUCTION

So far, we have considered Green's functions only for ordinary differential equations, i.e., in one-dimensional problems. However, Green's functions are useful for multidimensional spaces and general curvilinear coordinates as well as Cartesian coordinates. Thus an inverse operator can be defined for a partial differential operator as well as an ordinary differential operator. The Green's function will again satisfy the given differential equation with a (now multi-dimensional) δ function as the forcing function. Let's consider what is to be meant by a multidimensional δ function.

2. GREEN'S FUNCTIONS FOR MULTIDIMENSIONAL PROBLEMS IN CARTESIAN COORDINATES

To be specific, let's consider a three-dimensional space. Write

$$\int\int\int \phi(\xi, \eta, \zeta)\delta(x, y, z, \xi, \eta, \zeta)\, d\xi\, d\eta\, d\zeta = \phi(x, y, z)$$

for all continuous functions ϕ with compact support in the three-dimensional space. The $\phi(\xi, \eta, \zeta)$ are of course a set of testing functions. (Unless otherwise stated, the integrals are from $-\infty$ to ∞.) The $\delta(x, y, z, \xi, \eta, \zeta)$ can be written as the product $\delta(x-\xi)\delta(y-\eta)\delta(z-\zeta)$ since both expressions will give $\phi(x, y, z)$ on the right. Clearly we can generalize to n dimensions immediately.

If we assume a particular operator L now, such as $L = -\nabla^2 u$, the Green's function can be found from

$$-\nabla^2 G = \delta(x-\xi)\delta(y-\eta)\delta(z-\zeta)$$

G satisfies the homogeneous equation everywhere except at the source point ξ, η, ζ, i.e., a singularity exists at ξ, η, ζ. Assuming that f is a continuous function with compact support in the three-dimensional space,

$$u(x, y, z) = \iiint G(x, y, z, \xi, \eta, \zeta) f(\xi, \eta, \zeta) d\xi\, d\eta\, d\zeta$$

hence the solution to $Lu = f(x)$ is determinable. (The domain of L is the space of all functions $u(x, y, z)$ with piecewise continuous second derivatives such that $\iiint u^2\, dx\, dy\, dz$ and $\iiint (Lu)^2\, dx\, dy\, dz$ are finite.)

An n-dimensional δ function $\delta(x_1, \ldots, x_n) = \delta(x_1)\delta(x_2) \ldots \delta(x_n)$ since, if $\phi(x_1, \ldots, x_n)$ is a function in an n-dimensional domain of compact support, and, $d\tau$, the element of volume, is $dx_1 dx_2 \ldots dx_n$, multiplying *either* form of the δ function above by the testing function ϕ and integrating over the domain yields $\phi(0, \ldots, 0)$. If curvilinear coordinates are required, the element of volume involves the Jacobian of the transformation and the δ function as well as the operator L must be written in the new coordinates.

3. GREEN'S FUNCTIONS IN CURVILINEAR COORDINATES

It is not always desirable to use rectangular cartesian coordinates. It may be more convenient to specify boundary conditions, e.g., in spherical coordinates so we will consider briefly transformations to other curvilinear coordinate systems. Thus consider the transformation, for example, from x, y to plane polar coordinates r, θ by $x = r \cos \theta$, $y = r \sin \theta$ and the Jacobian $\partial(x, y)/\partial(r, \theta) = r$.

Since integration involving a testing function such as

$$\iint \phi(x, y)\delta(x-\xi)\delta(y-\eta)dxdy = \phi(\xi, \eta)$$

in the new coordinates would involve the volume element $r\, dr\, d\theta$, the δ functions $\delta(r-\alpha)\delta(\theta-\beta)$ would have to be divided by the Jacobian to serve their original function i.e., since the testing function becomes $\phi(r, \theta)$, then

$$\iint \phi(r, \theta)\delta(r-\alpha)\delta(\theta-\beta)\, dr\, d\theta = \phi(\alpha, \beta)$$

thus

$$\delta(x-\xi)\delta(y-\eta) = \frac{\delta(r-\alpha)\delta(\theta-\beta)}{r}$$

Suppose the source singularity is at the origin $x = y = 0$ and therefore at $r = 0$ in the new coördinates. The testing function $\phi(r, \theta)$ reduces at the origin to a function of r alone and the symbolic function must be only dependent on r. Hence $\delta(x)\delta(y)$ becomes

$$\frac{\delta(r)}{J\int_0^{2\pi} d\theta} = \frac{\delta(r)}{2\pi r}$$

Now consider spherical coordinates $x = r \sin\theta \cos\phi$, $y = r \sin\theta \sin\phi$, $z = r\cos\theta$. The Jacobian $J = r^2 \sin\theta$. The volume element is $r^2 \sin\theta\, dr d\theta d\phi$ so that

$$\delta(x-\xi)\delta(y-\eta)\delta(z-\zeta) = \frac{\delta(r-\alpha)\delta(\theta-\beta)\delta(\phi-\gamma)}{r^2 \sin\theta}$$

for a source point at ξ, η, ζ where r, θ, ϕ are all different from zero.

Suppose the source singularity lies along the z axis so $x = y = 0$, then in the new coordinates $\theta = 0$. We must have

$$\delta(x)\delta(y)\delta(z-\zeta) = \frac{\delta(r-\alpha)\delta(\theta)}{r^2 \sin\theta \int_0^{2\pi} d\phi}$$

$$= \frac{\delta(r-\alpha)\delta(\theta)}{2\pi r^2 \sin\theta}$$

Suppose that the singularity (source point) is at $x = y = z = 0$. In cartesian coordinates we are dealing with the multi-dimensional δ function

$$\delta(x)\delta(y)\delta(z)$$

In the spherical coordinates we have

$$\delta(r) \Big/ \int_0^{\pi}\int_0^{2\pi} J\, d\theta\, d\phi = \delta(r)/4\pi r^2$$

To solve the equation

$$-\nabla^2 G(x, y, z, \xi, \eta, \zeta) = \delta(x-\xi)\delta(y-\eta)\delta(z-\zeta)$$

we can transform coordinates so the source singularity at $x = \xi$, $y = \eta$, $z = \zeta$, is moved to the origin by writing $x' = x-\xi$, $y' = y-\eta$, $z' = z-\zeta$. Then letting ∇' indicate the Laplacian with respect to x', y', z',

$$\nabla'^2 G = -\delta(x')\delta(y')\delta(z')$$

Transforming to spherical coordinates

$$\frac{1}{r^2}\frac{\partial}{\partial r}\left[r^2\frac{\partial G}{\partial r}\right] + \frac{1}{r^2 \sin\theta}\frac{\partial}{\partial\theta}\left[\sin\theta\frac{\partial G}{\partial\theta}\right]$$

$$+ \frac{1}{r^2 \sin^2\theta}\frac{\partial^2 G}{\partial\phi^2} = \frac{\delta(r)}{4\pi r^2}$$

where the δ function on the right side has used the fact that the source singularity is at the origin because of the original transformation. Since the right side depends only on r, G depends only on r and

$$\frac{1}{r^2}\frac{\partial}{\partial r}\left[r^2\frac{\partial G(r)}{\partial r}\right] = \frac{\delta(r)}{4\pi r^2}$$

$$\frac{\partial}{\partial r}\left[r^2\frac{\partial G}{\partial r}\right] = \frac{\delta(r)}{4\pi}$$

4. PROPERTIES OF δ FUNCTIONS FOR MULTI-DIMENSIONAL CASE

If $x_1, x_2, \ldots, x_n$ and $\xi_1, \xi_2, \ldots, \xi_n$ are rectangular Cartesian coordinates of $\bar{x}$ and $\bar{\xi}$ in E^n then

$$\delta(\bar{x}-\bar{\xi}) = \delta(x_1-\xi_1)\delta(x_2-\xi_2)\ \ldots\ \delta(x_n-\xi_n)$$

If $\bar{x}$, $\bar{\xi} \in E^2$, and r, θ represents plane polar coordinates of x and ρ, σ are plane polar coordinates of $\bar{\xi}$ then

$$\delta(\bar{x}-\xi) = \frac{1}{r}\,\delta(r-\rho)\delta(\theta-\sigma)$$

If $\bar{x}$, $\bar{\xi} \in E^3$ and r, θ, z represent cylindrical coordinates for $\bar{x}$ and ρ, σ, ζ represent cylindrical coordinates for $\bar{\xi}$ then

$$\delta(\bar{x}-\bar{\xi}) = \frac{1}{r}\,\delta(r-\rho)\delta(\theta-\sigma)\delta(z-\zeta)$$

If $\bar{x}$, $\bar{\xi} \in E^3$, r,θ,ϕ represent spherical coordinates of $\bar{x}$ and ρ, σ, ζ represent spherical coordinates of $\bar{\xi}$ then

$$\delta(\bar{x}-\bar{\xi}) = \frac{1}{r^2 \sin\theta}\,\delta(r-\rho)\delta(\theta-\sigma)\delta(\phi-\zeta)$$

Chapter XVIII

THE Itô EQUATION AND A GENERAL STOCHASTIC MODEL FOR DYNAMICAL SYSTEMS

Activity in stochastic differential activities has been based heavily on the well known work of K. Itô who introduced stochastic integrals (Itô, 1942) to rigorously formulate the stochastic differential equation that determines Kolmogorov's diffusion process (Kolmogorov, 1931). The Itô stochastic differential equation $dx_t = f(x_t)\,dt + g(x_t)\,dw_t$ has long been used as a model for a dynamical system perturbed by white noise. We have assumed a basic probability space (Ω, F, μ) and that a generic point of Ω is represented by ω. The stochastic process $x = (x_t(\omega),\ 0 \le t < \infty)$ represents the "solution" of the above equation. We need not refer here to the considerable and well-documented work in stochastic calculus which has followed this pioneering work. More recently a new "input-output" point of view has developed in which (1) is regarded as a mapping taking an "input" w into an "output" which means any continuous process w is defined without the need for stochastic integration.

A similar idea has been available since 1961 and is most completely discussed in a recent book (Adomian, 1983). Rather than the white noise process, the stochastic process is a general process, characterized by its known statistical properties (expectation, correlation, etc.), which is the input to a stochastic dynamical system described by a differential equation $Ly = x$ where the linear differential operator L has one or more stochastic process coefficients, and is, therefore, a stochastic operator. The objective is to determine the expectation or mean $m_y(t)$ and the correlation $R_y(t_1, t_2)$ of the output $y(t)$ in terms of "stochastic Green's functions". This is analogous to the determination of output in terms of input for any ordinary differential equation in terms of an ordinary Green's function. An extended form considers $Fy = Ly + Ny = x$ where the added term Ny is nonlinear and can have deterministic and stochastic components $\dot{N}y$ and My. Some recent extensions consider also nonlinear terms $N(y, \dot{y}, \cdots, y^{(n)})$.

In modeling physical phenomena involving noise, we usually write

$$y(t) = y(0) + \int_0^t f(s, y(s))\, ds + \int_0^t g(s, y(s))\, dz(s) \quad (1)$$

where the last integral, with z the Wiener process, is the Itô integral. Equivalently, we can write

$$dy = f(t, y)\, dt + g(t, y)\, dz \quad (2)$$

which is called a "stochastic differential equàtion" but is understood in the sense of the integral equation (1). We can write u(t) = dz/dt since we do not intend to assume a white noise process and can therefore write

$$\frac{dy}{dt} = f(t, y) + g(t, y)u(t)$$

This is equivalent to the above form $Fy = x$ in which we consider x as the "input" rather than u(t). Imagine, e.g., a physical system (linear or nonlinear) or "stochastic filter" with input x(t, ω) and output y(t, ω). The system is described by $y = Hx$ or by the differential equation $Fy = x$ whose "solution" $y = F^{-1}x \equiv Hx$. The solution however, is only in a statistical sense, i.e., we ask for a statistical description of y in terms of a statistical description of x and a stochastic Green's function. Thus u(t) is a random coefficient in the differential operator so the equation is a stochastic differential equation, a differential equation with stochastic process coefficients describing a stochastic system with stochastic input x and allowing random initial conditions as well. This work has been considerably developed in recent years as a theory applicable to linear or nonlinear physical dynamical systems without restriction to white noise processes and the resulting stochastic calculus. Such systems are modeled by the differential equation $Fy = x$ where x is a stochastic process on a probability space (Ω, F, μ) for $0 \leq t < \infty$ and F is a nonlinear stochastic (differential) operator.

As McShane has pointed out, the nondifferentiability of the Wiener process is a mathematical property not a physical property (McShane, 1974). In the Itô integral $\int f\, dz$, the Wiener process z(t, ω) is not of bounded variation;

however, a Lipschitz condition on z is reasonable for physical processes so that the integral would be a well defined Riemann-Stieltjes integral. The physically reasonable and the mathematically tractable models have little in common. Hence we do not wish to use the Wiener process in stochastic differential equations modeling physical problems. Rather than the well known Itô form of the stochastic differential equation given by (2), we will write $\mathcal{F}y = x$ where x is a stochastic process and $\mathcal{F}$ is a nonlinear stochastic operator decomposable into linear and nonlinear parts and further decomposable into deterministic and random parts. This equation has been solved for polynomial, trigonometric, or exponential nonlinearities without restriction to Wiener processes. Coefficients of the differential operators are stochastic processes whose means, correlations, etc., are specified. "Solution" means finding the stochastic Green's function which yields similar statistical knowledge of the solution process. The question of a Lipschitz condition (e.g., in the equation $dy/dt = f(t, y)$) on $f(t, y)$ does not arise. The solution is not a Picard type approximation. Statistical separability is automatically achieved in terms involving the operator and the solution process without closure approximations.

The so-called "robustness" properties of stochastic calculus have been a matter of intensive investigation as a means of increasing the applicability of stochastic models as well as for interesting mathematical questions. The above stochastic operator point of view offers interesting and different mathematics for stochastic systems characterized by linear or nonlinear stochastic differential operator equations and Volterra integral equations. When combined with work of Kuznetsov, Stratonovich, and Tikhonov (1960), the stochastic operator approach allows determination of all the output statistical properties directly in terms of input statistical properties for real physical systems involving stochastic parameters.

Consider a stochastic process $x(t, \omega)$, $t \in T$, $\omega \in (\Omega, F, \mu)$, a p.s. It can be expanded in the form:

$$z(t, \omega) = m_z(t) + \sum_{\nu} v_{\nu}(\omega)\phi_{\nu}(t) \tag{3}$$

where the $v_{\nu}(\omega)$ are uncorrelated random variables with zero expected values and functions $\phi_{\nu}(t)$ which are deterministic functions of time. Each individual term $v_{\nu}\phi_{\nu}$ is referred to by Pugachev (1965) as an *elementary random function*. Such expansions are canonical expansions of a stochastic process or Karhunen-Loeve expansions whose value lies in the fact that linear operations on random functions can thereby be reduced to operations over the non-random functions $\phi_\nu(t)$ and therefore to ordinary operations of calculus. Analogously, we can write:

$$z(t, \omega) = m_z(t) + \int w(\lambda)\phi(t, \lambda)\, d\lambda \tag{4}$$

where $w(\lambda)$ is a white noise variation of the parameter λ and $\phi(t, \lambda)$ is a deterministic function with parameter t and argument λ. The integral plays the role of the sum in (3) and each (infinitely small)(uncorrelated) elementary random function is $w(\lambda)\phi(t, \lambda)\, d\lambda$. Thus (4) is a representation of a stochastic process in terms of white noise.

For the special case that $\phi(t, \lambda) = \delta(t-\lambda)$ then from (4)

$$z(t, \omega) = m_z(t) + \int w(\lambda)\delta(t-\lambda)\, d\lambda \tag{5}$$

$$= m_z(t) + w(t)$$

or $z(t, \omega) = w(t)$ if $m_z(t) = 0$. An Itô equation of the form $dy = f(y, t)\, dt + g(y, t)\, dw$ should be written for an arbitrary noise input as

$$dy = f(y, t)\, dt + g(y, t)\, dz \tag{6}$$

which is interpreted as an integral equation where z is not a martingale but an arbitrary process in general, and is the martingale if and only if $\phi(t, \lambda) = \delta(t-\lambda)$.

The Itô equation is now a *differential equation*

$$\frac{dy}{dt} = f(y, t) + g(y, t)\frac{dz}{dt} \tag{7}$$

$dz/dt = \alpha(t, \omega)$, a new process, hence

$$\frac{dy}{dt} = f(y, t) + g(y, t)\alpha(t, \omega) \tag{8}$$

or in Adomian's form

$$\mathit{F}y = x \tag{9}$$

where F is a linear or nonlinear stochastic operator (involving stochastic coefficients) and $x(t, \omega)$ is a stochastic process. (Adomian, 1980, 1983).

Assuming $m_z(t)$ is zero, the quantity dz/dt is defined by

$$\frac{dz}{dt} = \frac{d}{dt}\int w(\lambda)\phi(t, \lambda)\, d\lambda \tag{10}$$

from (4) or

$$\frac{dz}{dt} = \int w(\lambda)\phi'(t, \lambda)\, d\lambda \tag{11}$$

The Itô equation is for a system whose noise input is an elementary random function, i.e., a single "component" of a canonical expansion of a real stochastic process and only for the singular case where $\phi = \delta(t-\lambda)$ does it, and therefore the question of differentiation of w, ever arise.

Since z is an arbitrary stochastic process input, its covariance function $K_z(t, t')$ is assumed specified since solution of the differential equation means determination of the desired statistical properties of y directly in terms of similar knowledge of the input.

When $m_z(t) = 0$

$$z(t, \omega) = \int w(\lambda)\phi(t, \lambda)\, d\lambda \tag{12}$$

is a transformation of the white noise into z with the deterministic kernel $\phi(t, \lambda)$.

Then

$$K_z(t, t') = \int G(\lambda)\phi(t, \lambda)\overset{*}{\phi}(t', \lambda)\, d\lambda \tag{13}$$

where G(t) is the intensity of the white noise.

BIBLIOGRAPHY

Adomian, G., 'Stochastic Systems Analysis', Applied Stochastic Processes, Academic Press, 1980.

Adomian, G, Stochastic Systems, Academic Press, 1983.

Itô, K., 'Differential Equations Determining Markov Processes', Zenkoku Shijo Sugaku Danwakai (1942), 1077 1352-1400 (in Japanese).

Kolmogorov, A., 'Uber Analytische Methoden in der Wahrscheinlichkeitsrechnung', Math. Ann. 104 (1931), 415-458.

Kuznetsov, P.I., R.L. Stratonovich, and V.I. Tikhonov, 'Quasimoment Functions in the Theory of Random Processes', Theor. Probability App. 5 (1960), 80-97.
McShane, E.J., Stochastic Calculus and Stochastic Models, Academic Press, 1974.
Pugachev, V.S., Theory of Random Functions, Pergamon Press, 1965.

Chapter XIX

NONLINEAR PARTIAL DIFFERENTIAL EQUATIONS AND THE DECOMPOSITION METHOD

We have seen that solution of partial differential equations with multidimensional Green's functions is relatively involved. Even ordinary differential equations can have quite complicated Green's functions. As a result the integral

$\int G(x, \xi)f(\xi)\, d\xi$ or in the three dimensional case

$\int\int\int G\, f(\xi, \eta, \zeta)\, d\xi\, d\eta\, d\zeta$ can be very difficult to compute.

The decomposition technique of Adomian provides much more computable solutions as a generalization of the method of solution for ordinary differential equations even when strongly nonlinear and/or strongly stochastic. The basic technique is fully discussed by G. Adomian (Academic Press, 1983) and this chapter presents some further generalizations.

The method solves equations of the form $Fy = x$ where F may be a linear or nonlinear differential or partial differential operator. The operator may be deterministic or it may be a *stochastic operator*. F is not necessarily a differential operator. Forthcoming work of Adomian and his co-workers will deal with algebraic equations (polynomial or transcendental), difference equations, delay equations, etc.

We now proceed directly to the solution of general partial differential equations. Current developments in mathematical physics, energy problems, and other areas have given new impetus to research on nonlinear partial differential equations and linearization techniques. Unfortunately, such techniques which assume essentially that a nonlinear system is "almost linear" often have little physical justification. It has become vital not only to mathematics but to the areas of application, that new approaches be found. Fluid mechanics, soliton physics, quantum field theory, and nonlinear evolution equations are only a few of the areas which can benefit from more general methods. In some areas, such as turbulence studies, and, quite possibly, many or all of the other areas, once possibilities are recognized, stochasticity is also a significant factor in the actual physical behavior. In this section, we discuss further generalization of the decomposition method of Adomian (1981, 1983). This method is global in scope in that it will solve accurately operator equations involving differential operators describing general dynamical systems which may be nonlinear and stochastic. Such systems lead to nonlinear stochastic differential equations, systems of equations, or partial differential equations. Deterministic nonlinear equations can be considered as a special case where stochasticity vanishes (just as linear is a trivial special case of nonlinear).

Adomian's approach to these problems began with linear *stochastic operator* equations and since 1976 has evolved rapidly to nonlinear stochastic operator equations. Consider for example

$$\frac{\partial u}{\partial t} + a(t, x)\frac{\partial u}{\partial x} + b(t, x)\frac{\partial^2 u}{\partial x^2} = g(t, x)$$

which can be rewritten in terms of operators as

$$L_t u + L_x u = g(t, x)$$

where $L_t = (\partial/\partial t)$ and $L_x = a(\partial/\partial x) + b(\partial^2/\partial x^2)$. A similar *stochastic* equation

$$\frac{\partial u(x, t, \omega)}{\partial t} + A(x, t)u(x, t, \omega) + B(x, t, \omega)u(x, t, \omega)$$

$$= f(x, t, \omega)$$

where $\omega \in (\Omega, F, \mu)$, a probability space, f is a stochastic process, A is a deterministic coefficient but B is a stochastic process coefficient, can similarly be written

$$Lu + \mathcal{R}u = f$$

where $L = (\partial/\partial t) + A(x, t)$ is a deterministic operator and $\mathcal{R} = B$ is a stochastic operator, or

$$\mathcal{L}u = f$$

where $\mathcal{L}$ is a stochastic operator with deterministic and random parts, L being $\langle \mathcal{L} \rangle$ if $\mathcal{R}$ is zero-mean.

Let us consider then the operator equation

$$\mathcal{F}u = g$$

where F represents a differential operator which may be ordinary or partial, linear or nonlinear, deterministic or stochastic. We suppose F has linear and nonlinear parts, i.e., $Fu = Lu + Nu$ where L is a linear (stochastic) operator and N is a nonlinear (stochastic) operator - the italic letter indicates stochasticity. We may, of course, have a nonlinear term which depends upon derivatives of u as well as u but this will be discussed in other publications.

Since L may have deterministic and stochastic components, let $L = \mathrm{L} + R$ where conveniently $\mathrm{L} = \langle L \rangle$ and $R = L - \mathrm{L}$. This is not a limitation on the method but a convenience in explanation. It is necessary that L be invertible. If the above choice makes this difficult, we choose a simpler L and let R incorporate the remainder. Let $Nu = \mathrm{N}u + Mu$ where Nu indicates a deterministic part and Mu indicates a stochastic nonlinear term.

F may involve derivatives with respect to x,y,z,t or mixed derivatives. To avoid difficulties in notation which tend to obscure rather than clarify, we will assume the same probability space for each process and let $\mathrm{L}u = \mathrm{L}_x+\mathrm{L}_y+\mathrm{L}_z+\mathrm{L}_t$ where the operators indicate quantities like $\partial^2/\partial x^2$, $\partial/\partial y$, etc., but, for now, no mixed derivatives. Similarly, R is written as $R_x+R_y+R_z+R_t$. Mixed derivatives and product nonlinearities such as $u^2u'^3$, uu'', $f(u, u', \ldots, u^{(m)}$ can also be handled as shown elswhere (Adomian, 1983).

A simple Langevin equation is written

$$\mathrm{L}u = g$$

where $\mathrm{L} = (d/dt) + \beta$ and g is a white noise process. Langevin equations as used for modeling complex nonlinear phenomena in physics having the form $\dot{\psi} = f(\Psi) + \xi$ are represented by $\mathrm{L}u + \mathrm{N}u = g$, however, we will not make any Markovian or white noise restrictions. All processes will be physical processes without restriction to being Gaussian or stationary. In the KdV equation, for example, Fu would become $\mathrm{L}_tu+\mathrm{L}_xu+\mathrm{N}u$ where Nu is of the form uu_x (again a product nonlinearity). In equations of the Satsuma-Kaup type for soliton behavior, we have also such products uu_x, uu_{xxx}, u_xu_{xx}. Stochastic transport equations will fit nicely into our format since $\nabla^2 = \mathrm{L}_x+\mathrm{L}_y+\mathrm{L}_z$ and stochastic behavior in coefficients or inputs are easily included. For example, instead of

$Ly(\bar{r}, t) = \xi(\bar{r}, t, \omega)$ where ξ is a random source and $L = (\partial/\partial t) - D\nabla^2$ or $(\partial/\partial t) - \Lambda_{xyz}$, we can include nonlinear terms or stochastic behavior in the operator. In soliton equations such as the sine-Gordon equation, we have $L_t + L_x + N(u)$ where $N(u)$ includes the trigonometric nonlinearities. We can allow trigonometric, polynomial, exponential, or products, sums of products, etc. Or, as in the LAX theorem. $N(u) = f(u, u_x, u_{xx}, \dots)$.

Let us consider the general form:

$$[\mathcal{L}_x + \mathcal{L}_y + \mathcal{L}_z + \mathcal{L}_t]u + \mathcal{N}u = g(x, y, z, t) \quad (1)$$

where $\mathcal{N}(u)$ indicates any nonlinear terms possibly involving derivatives of u, products, etc. Since $\mathcal{L} = L + R$, we have

$$(L_x + L_y + L_z + L_t)u + (R_x + R_y + R_z + R_t)u + \mathcal{N}u = g \quad (2)$$

We emphasize that R_x, R_y, R_z, R_t, as well as the g, are not necessarily random. They may be random. Or, they may simply be a part of an entirely deterministic operator and be chosen only to make the remaining part easily invertible. We solve for L_xu, L_yu, L_zu, L_tu, in turn and then assuming inverses L_x^{-1}, L_y^{-1}, L_z^{-1}, L_t^{-1} exist

$$\begin{aligned}
L_x^{-1}L_xu &= L_x^{-1}[g-L_yu-L_zu-L_tu]-L_x^{-1}[R_x+R_y+R_z+R_t]u-L_x^{-1}\mathcal{N}u \\
L_y^{-1}L_yu &= L_y^{-1}[g-L_xu-L_zu-L_tu]-L_y^{-1}[R_x+R_y+R_z+R_t]u-L_y^{-1}\mathcal{N}u \\
L_z^{-1}L_zu &= L_z^{-1}[g-L_xu-L_yu-L_tu]-L_z^{-1}[R_x+R_y+R_z+R_t]u-L_z^{-1}\mathcal{N}u \\
L_t^{-1}L_tu &= L_t^{-1}[g-L_xu-L_yu-L_zu]-L_t^{-1}[R_x+R_y+R_z+R_t]u-L_t^{-1}\mathcal{N}u
\end{aligned} \quad (3)$$

A linear combination of these solutions is necessary. Therefore, adding and dividing by four, we write

$$u = u_0 - (1/4)\Big\{(L_x^{-1}L_y + L_y^{-1}L_x) + (L_x^{-1}L_z + L_z^{-1}L_x) +$$
$$+ (L_x^{-1}L_t + L_t^{-1}L_x) + (L_y^{-1}L_z + L_z^{-1}L_y) +$$
$$+ (L_t^{-1}L_y + L_y^{-1}L_t) + (L_z^{-1}L_t + L_t^{-1}L_z)\Big\} u - \qquad (4)$$
$$- (1/4)[L_x^{-1} + L_y^{-1} + L_z^{-1} + L_t^{-1}][R_x + R_y + R_z + R_t]u -$$
$$- (1/4)[L_x^{-1} + L_y^{-1} + L_z^{-1} + L_t^{-1}]Nu$$

where the term u_0 includes

$$(1/4)[L_x^{-1} + L_y^{-1} + L_z^{-1} + L_t^{-1}]g$$

as well as terms arising from the initial conditions which depend on the number of integrations involved in the inverse operators. Thus, $L_x^{-1}L_xu = u(x, y, z, t) - \Theta_x$ where $L_x\Theta_x = 0$. Thus, $L_x^{-1}L_xu = u(x, y, z, t) - u(0, y, z, t)$ if L_x involves a single differentiation. $L_x^{-1}L_xu = u(x, y, z, t) - u(0, y, z, t) - x\partial u(0, y, z, t)/\partial x$ for a second order operator, etc. Similarly $L_y^{-1}L_yu = u - \Theta_y$ where $\Theta_y = (x, 0, z, t)$ for a single differentiation in L_y, etc. Thus, we have the partial homogeneous solutions Θ_x, Θ_y, Θ_z, Θ_t analogous to the one-dimensional problems considered in earlier work where we wrote $L_t^{-1}L_tu(t) = \int_0^t (\partial u/\partial t)\, dt = u(t) - u(0)$ when $L_t \equiv d/dt$. Thus

$$u_0 = (1/4)[\Theta_x+\Theta_y+\Theta_z+\Theta_t] + (1/4)[L_x^{-1}+ L_y^{-1}+ L_z^{-1}+ L_t^{-1}]g \qquad (5)$$

We now write Nu, the nonlinear term, as $Nu = \Sigma_{n=0}^{\infty}A_n$ where A_n are the author's previously defined polynomials (Adomian, 1983) and assume our usual decomposition of u into $\Sigma_{n=0}^{\infty} u_n$, or equivalently, of $F^{-1}g$ into $\Sigma_{n=0}^{\infty} F_n^{-1}g$ to determine the individual components.

For m dimensional problems we can write in a more condensed form,

$$u = u_0 - (1/m) \sum_{j=i+1}^{m} \sum_{\substack{i=1 \\ (i \neq j)}}^{m-1} [L_{x_i}^{-1} L_{x_j} + L_{x_j}^{-1} L_{x_i}]u$$

$$- (1/m)[\sum_{i=1}^{m} L_{x_i}^{-1}][\sum_{i=1}^{m} R_{x_i}]u \tag{6}$$

$$- (1/m)[\sum_{i=1}^{m} L_{x_i}^{-1}] \sum_{n=0}^{\infty} A_n$$

where

$$u_0 = (1/m)\left\{\sum_{i=1}^{m} \Theta_i + \sum_{i=1}^{m} L_{x_i}^{-1}g\right\}.$$

Thus u_0 can be easily calculated. The following components of the decomposition follow in terms of u_0 without problems of statistical separability when stochasticity is involved. Thus,

$$u_1 = -(1/m) \sum_{j=i+1}^{m} \sum_{\substack{i=1 \\ (i \neq j)}}^{m-1} [L_{x_i}^{-1} L_{x_j} + L_{x_j}^{-1} L_{x_i}]u_0 -$$

$$- (1/m)[\sum_{i=1}^{m} L_{x_i}^{-1}][\sum_{i=1}^{m} R_{x_i}]u_0 -$$

$$- (1/m)[\sum_{i=1}^{m} L_{x_i}^{-1}]A_0$$

$\vdots$

$$u_n = -(1/m) \sum_{j=i+1}^{m} \sum_{\substack{i=1 \\ (i \neq j)}}^{m-1} [L_{x_i}^{-1} L_{x_j} + L_{x_j}^{-1} L_{x_i}] u_{n-1}$$

$$-(1/m)[\sum_{i=1}^{m} L_{x_i}^{-1}][\sum_{i=1}^{m} R_{x_i}] u_{n-1}$$

$$-(1/m)[\sum_{i=1}^{m} L_{x_i}^{-1}] A_{n-1}$$

and the complete solution is $u = \Sigma_{n=0}^{\infty} u_n$ and our n term approximation ϕ_n is given by $\phi_n = \Sigma_{i=0}^{n-1} u_i$.

For the particular problem here,

$$u_0 = (1/4)[\Theta_x + \Theta_y + \Theta_z + \Theta_t] +$$

$$+ (1/4)[L_x^{-1} + L_y^{-1} + L_z^{-1} + L_t^{-1}]g$$

$$\vdots$$

$$u_n = -(1/4)\Big\{(L_x^{-1}L_y + L_y^{-1}L_x) + (L_x^{-1}L_z + L_z^{-1}L_x) +$$

$$+ (L_x^{-1}L_t + L_t^{-1}L_x) + (L_y^{-1}L_z + L_z^{-1}L_y) +$$

$$+ (L_t^{-1}L_y + L_y^{-1}L_t) + (L_z^{-1}L_t + L_t^{-1}L_z)\Big\}u_{n-1} -$$

$$- (1/4)[L_x^{-1}+L_y^{-1}+L_z^{-1}+L_t^{-1}][R_x+R_y+R_z+R_t]u_{n-1} -$$

$$- (1/4)[L_x^{-1}+L_y^{-1}+L_z^{-1}+L_t^{-1}]A_{n-1}.$$

In the one-dimensional (m=1) case,

$$u_0 = \Theta_t + L_t^{-1}g$$

$$u_1 = -L_t^{-1}R_t u_0 - L_t^{-1}A_0 \quad \text{etc.}$$

For simplicity in writing, define

$$\frac{L_x^{-1} + L_y^{-1} + L_z^{-1} + L_t^{-1}}{4} \equiv L^{-1}$$

and

$$(1/4)[(L_x^{-1}L_y + L_y^{-1}L_x) + (L_x^{-1}L_z + L_z^{-1}L_x) +$$
$$+ (L_x^{-1}L_t + L_t^{-1}L_x) + (L_y^{-1}L_z + L_z^{-1}L_y) +$$
$$+ (L_t^{-1}L_y + L_y^{-1}L_t) + (L_z^{-1}L_t + (L_t^{-1}L_z)]$$
$$\equiv G$$

and $\frac{R_x + R_y + R_z + R_t}{4} = R$, (then for m = 4)

$$u = u_0 - Gu - L^{-1}Ru - L^{-1}Nu.$$

In a one-dimensional case, Gu vanishes and the 1/4 or 1/m factor is, of course, equal to one and we have

$$u = u_0 - L^{-1}Ru - L^{-1}Nu$$

which is precisely the basis of earlier solutions of nonlinear ordinary differential equations.

$$L^{-1} = (1/m) \sum_{i=1}^{m} L_{x_i}^{-1}$$

$$R = (1/m) \sum_{i=1}^{m} R_{x_i}$$

Now

$$u = u_0 - (1/m) \sum_{j=i+1}^{m} \sum_{i=1}^{m-1} [L_{x_i}^{-1} L_{x_j} + L_{x_j}^{-1} L_{x_i}]u -$$
$$- L^{-1}Ru - L^{-1}Nu \qquad (8)$$

which reduces to

$$u = u_0 - L^{-1}Ru - L^{-1}Nu$$

as previously written for ordinary differential equations when m = 1 since the second term vanishes.

1. PARAMETRIZATION AND THE A_n POLYNOMIALS

A parametrization of equation (8) into

$$u = u_0 - (1/m)\lambda \sum_{j=i+1}^{m} \sum_{i=1}^{m-1} [L_{x_i}^{-1}L_{x_j} + L_{x_j}^{-1}L_{x_i}]u$$

$$- \lambda L^{-1}Ru - \lambda L^{-1}Nu \tag{1}$$

and $u = F^{-1}g = \sum_n u_n$ into

$$u = \sum_n \lambda^n F_n^{-1} g = \sum_n \lambda^n u_n \tag{2}$$

has been *convenient* in determining the components of u and also in finding the A_n polynomials originally. We will later set $\lambda = 1$ so $u = \sum_n u_n$. *The λ is not a perturbation parameter.* It is simply an identifier helping us to collect terms in a way which will result in each u_i depending only on $u_{i-1}, u_{i-2}, \ldots, u_0$.

Now $N(u)$ is a nonlinear function and $u = u(\lambda)$. We assume $N(u)$ is analytic and write it as $\Sigma A_n \lambda^n$ if $N(u) = Nu$, i.e., if N is deterministic. (If the nonlinear stochastic term Mu appears, we simply carry that along or a second "stochastically analytic" expansion $\Sigma B_n \lambda^n$.)

Now (9) becomes

$$\sum_n \lambda^n F_n^{-1} g = u_0 - (1/m)\lambda \sum_{j=i+1}^{m} \sum_{i=1}^{m-1} [L_{x_i}^{-1}L_{x_j} + L_{x_j}^{-1}L_{x_i}] \sum_n \lambda^n F_n^{-1} g -$$

$$- \lambda L^{-1} R \sum_n \lambda^n F_n^{-1} g - \lambda L^{-1} \sum_n \lambda^n A_n.$$

Equating powers of λ

$$F_0^{-1}g = u_0$$

$$F_1^{-1}g = -(1/m) \sum_{j=i+1}^{m} \sum_{i=1}^{m-1} [L_{x_i}^{-1}L_{x_j} + L_{x_j}^{-1}L_{x_i}](F_0^{-1}g) -$$

$$- L^{-1}R(F_0^{-1}g) - L^{-1}A_0 \tag{4}$$

$$\vdots$$

$$F_n^{-1}g = -(1/m) \sum_{j=i+1}^{m} \sum_{i=1}^{m-1} [L_{x_i}^{-1}L_{x_j} + L_{x_j}^{-1}L_{x_i}](F_{n-1}^{-1}g) -$$

$$- L^{-1}R(F_{n-1}^{-1}g) - L^{-1}A_{n-1}.$$

Hence all terms are calculable. If there is both a deterministic and a stochastic term which is nonlinear, i.e., $\mathcal{N}u = Nu + \mathcal{M}u$ we have also $-L^{-1}B_{n-1}$ calculated the same way but involving randomness. If randomness is involved anywhere in any part of the equation, we will then calculate the statistical measures - the expectation and covariance of the solution process.

Thus each $F_{n+1}^{-1}g$ depends on $F_n^{-1}g$ and ultimately on $F_0^{-1}g$. Hence, F^{-1} the stochastic nonlinear inverse has been determined. The quantities A_n and B_n have been calculated for general classes of nonlinearities (Adomian, 1983) and explicit formulas have been developed. Their calculation is as simple as writing down a set of Hermite or Legendre polynomials. They depend, of course, on the particular nonlinearity.

If stochastic quantities are involved, the above series then involves processes and can be averaged for $\langle u \rangle$ or multiplied and averaged to form the correlation $\langle u(t_1)\overset{*}{u}(t_2)\rangle = R_u(t_1, t_2)$ as discussed in the author's previous works. Thus, the solution statistics (or statistical measures) are obtained when appropriate statistical knowledge of the random quantities is available.

Summarizing, we have decomposed the solution process for the output of a physical system into additive components - the first being the solution of a simplified linear deterministic system which takes account of initial conditions. Each of the other components is then found in terms of a *preceding* component and thus ultimately in terms of the first.

The usual statistical separability problems, leading to closure approximations, are eliminated with the reasonable assumption of statistical independence of the system *input* and the system itself! Quasimonochromaticity assumptions are unnecessary and processes can be assumed to be general physical processes rather than white noise. White noise is not a physical process. Physical inputs are neither unbounded nor do they have zero correlation times. In any event, the results can be obtained as a special case. If fluctuations are small, the results of perturbation theory are exactly obtained but again this is a special case, as are the diagrammatic methods of physicists. (Adomian, 1983).

Just as spectral spreading terms are lost by a quasi-monochromatic approximation when we have a random or scattering medium, and terms are also lost in the use of closure approximations, Boussinesq approximations, replacement of stochastic quantities by their expectations, etc., significant terms may be lost by the usual linearizations, unless of course, the behavior is actually close to linear.

One hopes, therefore, that physically more realistic and accurate results and predictions will be obtained in many physical problems by this method of solution, as well as interesting new mathematics from the study of such operators and relevant analysis.

Examples: Two-dimensional Partial Differential Equation: Consider the homogeneous partial differential equation

$$(L_x + L_y)\, u = 0$$

where $L_x = \partial/\partial x$ and $L_y = \partial/\partial y$ with $u(0, y) = -y$ and $u(x, 0) = x$. Solving for $L_x u$ and $L_y u$ separately

$$L_x u = -L_y u$$

$$L_y u = -L_x u$$

Then

$$L_x^{-1} L_x u = -L_x^{-1} L_y u \tag{5}$$

$$L_y^{-1} L_y u = -L_y^{-1} L_x u \tag{6}$$

The left side of (5) yields

$$L_x^{-1}L_x u = u - \phi_x \tag{7}$$

where $L_x\phi_x = 0$. This is easy to see. Write $L_x^{-1}L_x u$ as

$$\int_0^x \frac{\partial u}{\partial x}\, dx = u(x, y)\Big|_0^x = u(x, y) - u(0, y) = u - \phi_x .$$

The notation means merely that this arises from the L_x equation. ϕ_x is, of course, a function of y. Similarly (6) yields

$$L_y^{-1}L_y u = u - \phi_y \tag{8}$$

where $\phi_y = u(x, 0)$. ϕ_x and ϕ_y are *partial homogeneous solutions*. This is analogous to the one-dimensional solutions (Adomian, 1983) where we write $L_t^{-1}L_t u = \int_0^t (\partial u/\partial t)\, dt =$ $= u(t) - u(0)$. From (5), (7) we have

$$u - \phi_x = -L_x^{-1}L_y u \tag{9}$$

From (6), (8) we have

$$u - \phi_y = -L_y^{-1}L_x u \tag{10}$$

Adding and dividing by two,

$$u = (1/2)[\phi_x + \phi_y] - (1/2)[L_x^{-1}L_y + L_y^{-1}L_x]u \tag{11}$$

Using the decomposition method (Adomian, 1983), identify

$$u_0 = (1/2)[\phi_x + \phi_y] = (1/2)(x-y)$$

Consequently

$$u = u_0 - (1/2)[L_x^{-1}L_y + L_y^{-1}L_x][u_0 + u_1 + \ldots]$$

and we obtain immediately since $u = u_0 + u_1 + \ldots$

$$u_1 = -(1/2)[L_x^{-1}L_y + L_y^{-1}L_x]u_0$$

$$= -(1/2)[L_x^{-1}L_y + L_y^{-1}L_x](1/2)(\phi_x+\phi_y)$$

i.e., u_1 is calculated in terms of u_0. Similarly u_2 is calculated in terms of u_1, etc. The result is

$$u = \sum_{n=0}^{\infty} (-1)^n(1/2)^{n+1}[L_x^{-1}L_y + L_y^{-1}L_x]^n(\phi_x+\phi_y)$$

If we write $Lu = (L_x + L_y)u = 0$, i.e., $u(x, y)$ is the solution of the homogeneous equation $Lu = 0$, we have the partial homogeneous solutions ϕ_n and ϕ_y satisfying $L_x\phi_x = 0$ and $L_y\phi_y = 0$ and analogously the multidimensional homogeneous solution satisfying $Lu = 0$.

Thus we have calculated the multidimensional homogeneous solution in terms of one-dimensional, or partial homogeneous solutions.

Exercise: Calulate u_1, u_2, ... Show the result converges to $u = x-y$.

Thus the solution of the partial differential equation is $u(x, y) = x-y$. (Verify by substitution.) Sometimes of course we will not be able to sum the infinite series or see an algorithm. We can however always compute a number of terms. Usually half a dozen terms are sufficient. In this particular case we get accuracy within 1% by 7 terms. By 10 terms, we have 0.1% accuracy, i.e., the solution $u = 0.999\ (x-y)$.

Example: Four-dimensional Linear Partial Differential Equation: Consider the equation

$$(\partial u/\partial x) + (\partial u/\partial y) + (\partial u/\partial z) + (\partial u/\partial t) = 0.$$

Assume initial conditions

$$u(0, y, z, t) = \Theta_x = f_1(y, z, t) = -y + z - t$$

$$u(x, 0, z, t) = \Theta_y = f_2(x, z, t) = x + z - t$$

$$u(x, y, 0, t) = \Theta_z = f_3(x, y, t) = x - y - t$$

$$u(x, y, z, 0) = \Theta_t = f_4(x, y, z) = x - y + z$$

(We note u(0, 0, 0, 0) = 0.) In our usual notation we write

$$L_x + L_y + L_z + L_t = \mathcal{L}_{x,y,z,t}$$

and

$$\mathcal{L}u = 0.$$

By the decomposition method, u(x, y, z, t) is given by:

$$u = \sum_{n=0}^{\infty} (-1)^n \left(\frac{1}{4}\right)^n \left\{ \sum_{j=i+1}^{4} \sum_{i=1}^{3} (L_{x_i}^{-1} L_{x_j} + L_{x_j}^{-1} L_{x_i}) \right\}^n \cdot \left(\frac{1}{4}\right)\left(\sum_{m=1}^{4} \Theta_m \right).$$

The last sum $\sum_{m=1}^{4} \Theta_m = 3x-3y+3z-3t$ and the double summation within the curly brackets is

$$(L_x^{-1}L_y + L_y^{-1}L_x) + (L_x^{-1}L_z + L_z^{-1}L_x) + (L_x^{-1}L_t + L_t^{-1}L_x) +$$

$$+ (L_y^{-1}L_z + L_z^{-1}L_y) + (L_y^{-1}L_t + L_t^{-1}L_y) + (L_z^{-1}L_t + L_t^{-1}L_z).$$

We seek to find our approximate solution ϕ_n

$$\phi_n = u_0 + u_1 + \cdots + u_{n-1}$$

for as high an n as we wish to calculate it. We have

$$u_0 = \frac{1}{4} \sum_{m=1}^{4} \Theta_m = \frac{1}{4} (3x-3y+3z-3t)$$

$$= \left(\frac{3}{4}\right)(x-y+z-t)$$

$$u_1 = -\frac{3}{16} \{ \cdot \}(x-y+z-t)$$

where the double summation in the curly brackets $\{\cdot\}$ is easily evaluated. Thus

$$(L_x^{-1}L_y)(x-y+z-t)$$

$$= L_x^{-1}(-1) = \int_0^x (-1)\, dx = -x$$

$$(L_y^{-1}L_x)(x-y+z-t)$$

$$= (L_y^{-1})(1) = \int_0^y dy = y$$

Hence $(L_x^{-1}L_y+L_y^{-1}L_x)(x-y+z-t) = -x+y$. Similarly evaluating the other five quantities in the double summation,

$$u_1 = -\frac{3}{16}[-x+y+x+z-x+t+y-z-t-y-z+t]$$

$$u_1 = -\frac{3}{16}[-x+y-z+t]$$

$$= \frac{3}{16}[x-y+z-t]$$

(Thus a two term approximation $\phi_2 = u_0+u_1$ would be given by $\phi_2 = \frac{3}{4}(x-y+z-t) + \frac{3}{16}(x-y+z-t) = \frac{15}{16}(x-y+z-t)$.)

Continuing in the same manner, we get

$$u_n = \frac{3}{4^{n+1}}(x-y+z-t)$$

and since $u = \sum_{n=0}^{\infty} u_n$

$$u = k(x-y+z-t)$$

where

$$k = \sum_{n=0}^{\infty} 3/4^{n+1}$$

The series for k is given by $3(\frac{1}{4^1} + \frac{1}{4^2} + \frac{1}{4^3} + \frac{1}{4^4} + \frac{1}{4^5} + \frac{1}{4^6} + \cdots)$ or $3(\frac{1}{2^2} + \frac{1}{2^4} + \frac{1}{2^6} + \frac{1}{2^8} + \cdots) = \frac{3}{4}(\frac{1}{2^0} + \frac{1}{2^2} + \frac{1}{2^4} + \cdots)$.

The series for $k = \frac{3}{4}\left(\frac{1}{2^0} + \frac{1}{2^2} + \frac{1}{2^4} + \cdots\right) = 0.75 + 0.1875 +$ $+ 0.046875 + 0.01171875 + 0.0029296815 + 0.0007324219 + \cdots$. The sum of the first 5 terms is 0.9990 and for six terms, the sum is 0.9998 to four places. Thus the correct solution is

$$u = x-y+z-t$$

to within 0.02% error. Of course it's trivial to verify that this answer is indeed the correct solution.

Example: Nonlinear P.D.E.:

Let $\nabla^2 = (\partial^2/\partial x^2) + (\partial^2/\partial y^2) + (\partial^2/\partial z^2) = L_x+L_y+L_z$

(where the notation L_x symbolizes a linear (deterministic) differential operator $\partial^2/\partial x^2$, etc.) and N(p), or Np for convenience, symbolizes the nonlinear (deterministic) operator acting on p.

We will consider the nonlinear equation $\nabla^2 p = k^2 \sinh p$ which we write as[1]

$$(L_x + L_y + L_z)p = k^2 Np$$

where Np = sinh p. Hence

$$L_x p = k^2 Np - L_y p - L_z p$$
$$L_y p = k^2 Np - L_x p - L_z p$$
$$L_z p = k^2 Np - L_x p - L_y p.$$

Assuming the inverses L_x^{-1}, L_y^{-1}, L_z^{-1} exist[2]

$$L_x^{-1} L_x p = L_x^{-1} k^2 Np - L_x^{-1} L_y p - L_x^{-1} L_z p$$

$$L_y^{-1} L_y p = L_y^{-1} k^2 Np - L_y^{-1} L_x p - L_y^{-1} L_z p \qquad (12)$$

$$L_z^{-1} L_z p = L_z^{-1} k^2 Np - L_z^{-1} L_x p - L_z^{-1} L_y p.$$

But

$$
\begin{aligned}
L_x^{-1}L_xp &= p - p(0, y, z) - x\,\frac{\partial p}{\partial x}(0, y, z)\\
L_y^{-1}L_yp &= 0 - p(x, 0, z) - y\,\frac{\partial p}{\partial y}(x, 0, z)\\
L_z^{-1}L_zp &= p - p(x, y, 0) - z\,\frac{\partial p}{\partial z}(x, y, 0)
\end{aligned}
\tag{13}
$$

Consequently, summing (12) with the substitution (13), and dividing by three

$$
\begin{aligned}
p = \frac{1}{3}\,[&p(0, y, z) + x\,\frac{\partial p}{\partial x}(0, y, z) +\\
&+ p(x, 0, z) + y\,\frac{\partial p}{\partial y}(x, 0, z) +\\
&+ p(x, y, 0) + z\,\frac{\partial p}{\partial z}(x, y, 0)] +\\
&+ \frac{1}{3}\,[(L_x^{-1}L_y + L_y^{-1}L_z) + (L_x^{-1}L_z + L_z^{-1}L_x) +\\
&+ (L_y^{-1}L_z + L_x^{-1}L_y)]p +\\
&+ \frac{1}{3}\,[L_x^{-1} + L_y^{-1} + L_z^{-1}]\,\Sigma\,A_n
\end{aligned}
\tag{14}
$$

where N(u) has been replaced by Adomian's A_n polynomials. We take immediately the first term of our approximation

$$
\begin{aligned}
p_0 = \frac{1}{3}[&p(0, y, z) + x\,\frac{\partial p}{\partial x}(0, y, z) +\\
&+ p(x, 0, z) + y\,\frac{\partial p}{\partial y}(x, 0, z) +\\
&+ p(x, y, 0) + z\,\frac{\partial p}{\partial z}(x, y, 0)]
\end{aligned}
\tag{15}
$$

and following terms given by

$$
\begin{aligned}
p_1 = \frac{1}{3}[&(L_x^{-1}L_y + L_y^{-1}L_z) + (L_x^{-1}L_z + L_x^{-1}L_x) +\\
&+ (L_y^{-1}L_z + L_z^{-1}L_y)]p_0 + \frac{1}{3}[L_x^{-1} + L_y^{-1} + L_z^{-1}]A_0
\end{aligned}
\tag{16}
$$

$$\vdots$$

$$p_n = \frac{1}{3}[(L_x^{-1}L_y + L_y^{-1}L_z) + (L_x^{-1}L_z + L_z^{-1}L_x) +$$

$$+ (L_y^{-1}L_z + L_z^{-1}L_y)p_{n-1} +$$

$$+ \frac{1}{3}[L_x^{-1} + L_y^{-1} + L_z^{-1}]A_{n-1}.$$

Now we must evaluate the A_n which have been discussed adequately elsewhere (Adomian, 1983). We use the fact that $(d/dx)(\sinh x) = \cosh x$ and $(d/dx) \cosh x = \sinh x$ and the formulas for A_n given in the references.

$$H_0(p_0) = \sinh p_0$$

$$H_1(p_0) = \cosh p_0$$

$$H_2(p_0) = \sinh p_0$$

$$\vdots$$

i.e.,

$$H_n(p_0) = \sinh p_0 \quad \text{for even } n$$

$$H_n(p_0) = \cosh p_0 \quad \text{for odd } n.$$

Then

$$A_0 = \sinh p_0$$

$$A_1 = p_1 \cosh p_0$$

$$A_2 = p_2 \cosh p_0 + \frac{1}{2} p_1^2 \sinh p_0 \tag{17}$$

$$A_3 = p_3 \cosh p_0 + p_1 p_2 \sinh p_0 + \frac{1}{6} p_1^3 \cosh p_0$$

$$A_4 = p_4 \cosh p_0 + [\frac{1}{2} p_2^2 + p_1 p_3] \sinh p_0$$
$$+ \frac{1}{2} p_1^2 p_2 \cosh p_0 + \frac{1}{24} p_1^4 \sinh p_0$$

etc.

The A_n are easily written down by the procedures given in the references for as many terms as desired. From (17), we have the complete decomposition $p = \Sigma_{n=0}^{\infty} p_i$ and hence the solution.

Many more examples and applications to problems of physics and engineering will appear in forthcoming publications.

Example: Nonlinear Hyperbolic Initial-Value Problem: Consider the partial differential equation

$$\frac{\partial^2 u}{\partial x \partial y} + g(x, y) f(u) = 0 \tag{18}$$

$$u(x, 0) = \phi(x)$$

$$u(0, y) = \Psi(y)$$

This equation has been considered by Hsiang and Kwong (1982) for $\{(x, y) \in R^2, x \geq 0, y \geq 0\}$ with some appropriate smoothness conditions on g, f, ϕ, Ψ to guarantee existence, uniqueness, and continuous dependence on the coefficient g and the initial conditions. We propose here an approximate solution using the methods of Adomian. Let $L = \partial^2/\partial x \partial y$ and write (18) as

$$Lu = g\, f(u) = 0$$

or

$$Lu = -g\, f(u)$$

L^{-1} involves integration with respect to x and y from 0 to x and 0 to y respectively. Therefore

$$L^{-1}Lu = \int_0^x dx \int_0^y dy\ u_{xy}(x, y)$$

$$= u(x, y) - u(0, y) - u(x, 0) + u(0, 0)$$

Consequently

$$u(x, y) = u(0, y) + x\ u(x, 0) + u(0, 0) - L^{-1}\ g\ f(u)$$

Let $u(0, y) - u(x, 0) = u_0$ and decompose u into $\Sigma_{n=0}^{\infty} u_n$ with u_0 as given above. The components u_1, u_2, ... are to be determined. Using the previously developed methods, we parametrize the equation and the sum of the u_n. Then

$$\sum_{n=0}^{\infty} \lambda^n u_n = u_0 - \lambda L^{-1} g\ f(u)$$

The nonlinear term f(u) (or Nu in our generic equation) is expanded in terms of the A_n polynomials. Then

$$\Sigma\lambda^n u_n = u_0 - \lambda L^{-1} g(x, y)\Sigma\lambda^n A_n$$

Since u_0 is known,

$$u_1 = -L^{-1}\ g(x, y)A_0$$

$$u_2 = -L^{-1}\ g(x, y)A_1$$

$$\vdots$$

$$u_n = -L^{-1}\ g(x, y)A_{n-1}$$

The parameter λ is no longer necessary; it was merely a convenient way of obtaining the u_i for i=1, 2, ... The complete solution is given by

$$u = u_0 - \sum_{n=1}^{\infty} L^{-1} g(x, y)A_n$$

where L^{-1} is the previously given double integration operator, $g(x, y)$ is known, and only the A_n remain to be calculated. These of course depend upon the particular nonlinearity and are determinable as discussed in the referenced works and briefly in the following section.

2. INVERSES FOR NON-SIMPLE DIFFERENTIAL OPERATORS

In solving partial differential equations, we notice terms like $L_x^{-1}L_y + L_y^{-1}L_x$ occur in our solution algorithm. When L_x, L_y are simple differential operators, e.g., like $L_x = \partial^2/\partial x^2$, $L_y = \partial/\partial y$ then of course the inverse operators are immediately determined. Suppose however we have $L_x = \Sigma_{n=0}^{N_x} \alpha_n(x, y)\, \partial^n/\partial x^n$ and $L_y = \Sigma_{n=0}^{N_y} \beta_n(x, y)\, \partial^n/\partial y^n$ where N_x indicates the order of the L_x operator and N_y indicates the order of the L_y operator. We will suppose the highest order terms have coefficients of unity, i.e., $\alpha_{N_x} = \beta_{N_y} = 1$. Now the highest order terms are simple $\partial^{N_x}/\partial x^{N_x} = L_{x_0}$ and $\partial^{N_y}/\partial x^{N_y} = L_{y_0}$. In other words, L_{x_0}, L_{y_0} are simple differential operators such as we were using to get an easily inverted operator. Then

$$L_x = L_{x_0} + \sum_{n=0}^{N_x-1} \alpha_n(x, y)\partial^n/\partial x^n$$

$$L_y = L_{y_0} - \sum_{n=0}^{N_y-1} \beta_n(x, y)\partial^n/\partial y^n$$

The inverses can now be found by the decomposition method itself and are given by

$$L_x^{-1} = \sum_{n=0}^{\infty} (-1)^n (L_{x_0}^{-1} L_{x_1})^n L_{x_0}^{-1}$$

where

$$L_{x_1} = \sum_{n=0}^{N_x - 1} \alpha_n(x, y)\, \partial^n/\partial x^n$$

and L_y^{-1} is similarly determined.

3. MULTIDIMENSIONAL GREEN'S FUNCTIONS BY DECOMPOSITION METHOD

Consider the equation $\mathcal{L}u = g(x_1, \ldots, x_n)$ where $\mathcal{L} = \Sigma_{i=1}^{n} L_{x_i}$. We would like to write

$$\mathcal{L}G = \delta = \sum_{i=1}^{n} \delta(x_i - \xi_i)$$

in analogy with one-dimensional problems. We have noted in our solutions of nonlinear partial differential equations that the algorithm for the solution involves one-dimensional inverses, i.e., a multidimensional inverse operator or, equivalently, a Green's function is essentially being determined in terms of one-dimensional inverses which are much easier to determine. Let's consider this further as follows. We can write for a one-dimensional case,

$$L_x^{-1} v = \int_0^x G_1(x, \xi) v(\xi, y)\, d\xi$$

$$L_y^{-1} v = \int_0^y G_2(y, \eta) v(x, \eta)\, d\eta$$

calling G_1, G_2 "partial Green's functions" for the equation $(L_x + L_y)u = g(x, y)$. For this two-dimensional equation, G must satisfy

$$(L_x + L_y)G = \delta(x-\xi)\delta(y-\eta) = \delta_{x,\, y}$$

Now

$$L_x G = \delta_{x,\ y} - L_y\ G$$

$$L_y G = \delta_{x,\ y} - L_x\ G$$

from which

$$G = L_x^{-1}\delta_{x,\ y} - L_x^{-1}L_y G$$

$$G = L_y^{-1}\delta_{x,\ y} - L_y^{-1}L_x G$$

Consequently

$$G = (1/2)[L_x^{-1} + L_y^{-1}]\delta_{x,\ y} - (1/2)[L_x^{-1}L_y + L_y^{-1}L_x]G$$

which we can solve by the Adomian decomposition writing

$$G(x,\ \xi;\ y,\ \eta) = \sum_{n=0}^{\infty} G_n$$

where

$$G_0 = (1/2)[L_x^{-1} + L_y^{-1}]\delta(x-\xi)\delta(y-\eta)$$

$$G_1 = -(1/2)[L_x^{-1}L_y + L_y^{-1}L_x]G_0$$

$$\vdots$$

so that

$$G(x,\ \xi;\ y,\ \eta) = \sum_{n=0}^{\infty} (-1)^n(\tfrac{1}{2})^{n+1}[L_x^{-1}L_y+L_y^{-1}L_x]^n[L_x^{-1}+L_y^{-1}]\ \delta(x-\xi)\delta(y-\eta)$$

Now

$$u(x,\ y) = \int_0^x d\xi \int_0^y d\eta\Big\{\sum_{n=0}^{\infty} (-1)^n(\tfrac{1}{2})^{n+1}[L_x^{-1}L_y+L_y^{-1}L_x]^n[L_x^{-1}+L_x^{-1}]$$

$$\cdot\ \delta(x-\xi)\delta(y-\eta)\Big\}g(\xi,\ \eta)$$

i.e.,

$$u(x, y) = \int_0^x d\xi \int_0^y d\eta\ G(x, \xi; y, \eta) g(\xi, \eta)$$

is the representation of $L^{-1} g(x, y) = [L_x + L_y]^{-1} g(x, y)$

4. RELATIONSHIPS BETWEEN GREEN'S FUNCTIONS AND THE DECOMPOSITION METHOD FOR PARTIAL DIFFERENTIAL EQUATIONS

Consider an equation such as $\nabla^2 u(x, y) = g(x, y)$ where $x, y \in R^2$, $x \in [x_1, x_2]$, and $y \in [y_1, y_2]$ given specific boundary conditions for the intervals $[x_1, x_2]$ and $[y_1, y_2]$. We have

$$\frac{\partial^2 u}{\partial x^2} + \frac{\partial^2 u}{\partial y^2} = g(x, y)$$

which we write

$$(L_x + L_y) u(x, y) = g(x, y)$$

by letting $L_x = \partial^2/\partial x^2$ and $L_y = \partial^2/\partial y^2$. Now writing

$$L_x u = g - L_y u$$

$$L_y u = g - L_x u$$

and

$$L_x^{-1} L_x u = L_x^{-1} g - L_x^{-1} L_y u$$

$$L_y^{-1} L_y u = L_y^{-1} g - L_y^{-1} L_x u$$

$$L_x^{-1} L_x u = L_x^{-1} [\partial^2 u/\partial x^2] = u - \alpha_1 - \alpha_2 x \text{ and } L_y^{-1} L_y u = u - \beta_1 - \beta_2 y.$$

Consequently

$$u = \alpha_1 + x\alpha_2 + L_x^{-1}g - L_x^{-1}L_y u$$

$$u = \beta_1 + y\beta_2 + L_y^{-1}g - L_y^{-1}L_x u$$

from which

$$u = (1/2)\{[\alpha_1 + x\alpha_2 + \beta_1 + y\beta_2] +$$

$$+ [L_x^{-1} + L_y^{-1}]g - [L_x^{-1}L_y + L_y^{-1}L_x]\}u$$

Identifying

$$u_0 = (1/2)[\alpha_1 + x\alpha_2 + \beta_1 + y\beta_2] + (1/2)[L_x^{-1} + L_y^{-1}]g$$

Now all u_n for $n \geq 1$ are determined by the equation

$$u_n = -(1/2)[L_x^{-1}L_y + L_y^{-1}L_x]u_{n-1}$$

and the solution $u = u_0 + \Sigma_{n=1}^{\infty} u_n$ where u must satisfy the requirements imposed by the boundary conditions. It has been suggested by Adomian that the Green's function can similarly be found and that the method may be valuable when an ordinary Green's function is difficult to find or, because of its complexity, not useful for computation. If $G(x, y; \xi, \eta)$ is the Green's function, we can write

$$u = \int_{x_1}^{x_2} d\xi \int_{y_1}^{y_2} d\eta \; G(x, y; \xi, \eta)g(\xi, \eta)$$

for an arbitrary inhomogeneous term. It is also true that u_0 in the decomposition of the solution is given by

$$u_0 = \int_{x_1}^{x_2} d\xi \int_{y_1}^{y_2} d\eta \; G_0(x, y; \xi, \eta)g(\xi, \eta)$$

where G, or G_0, must satisfy the boundary conditions imposed on u. With reference to our previous discussion on Green's functions for partial differential equations, $G(x, y; \xi, \eta)$ must satisfy

$$(L_x + L_y)G(x, y; \xi, \eta) = \delta(x-\xi)\delta(y-\eta)$$

If we use the decomposition procedure to solve for $G = \sum_{n=0}^{\infty} G_n(x, y; \xi, \eta)$, we solve for $L_x G$ and $L_y G$ in turn, operate on the equations with the appropriate inverse operators as before, add and divide by two just as before. Then we find that

$$G(x, y; \xi, \eta) = (1/2)[\alpha_1 + x\alpha_2 + \beta_1 + y\beta_2]$$

$$+ (1/2)[L_x^{-1} + L_y^{-1}]\delta(x-\xi)\delta(y-\eta) +$$

$$+ (1/2)[L_x^{-1}L_y + L_y^{-1}L_x]G(x, y; \xi, \eta)$$

Identify as G_0 the terms above

$$(1/2)[\alpha_1 + x\alpha_2 + \beta_1 + y\beta_2]$$

$$+ (1/2)[L_x^{-1} + L_y^{-1}]\delta(x-\xi)\delta(y-\eta)$$

Calculating the second part above by doing the simple integrations represented by $L_y^{-1}\delta(x-\xi)\delta(y-\eta)$ and $L_x^{-1}\delta(x-\xi)\delta(y-\eta)$ and adding, we have

$$G_0 = (1/2)[(x-\xi)H(x-\xi)\delta(y-\eta)+(y-\eta)H(y-\eta)\delta(x-\xi)] +$$

$$+ (1/2)[\alpha_1(\xi)+x\alpha_2(\xi)+\beta_1(\eta)+y\beta_2(\eta)]$$

The α's and β's are evaluated by use of the boundary conditions imposed on u. Thus G_0 can be used to determine u_0 and proceeding to get u, or, alternately, to get the entire Green's function G from which u is determined.

Boundary conditions can be specified as requirements on u, u' on the boundaries of the intervals of interest

$$\beta_1(y) = k_{11}u'(x_1, y) + k_{12}\, u(x_1, y)$$

$$\beta_2(y) = k_{21}u'(x_2, y) + k_{22}\, u(x_2, y)$$

$$\beta_3(x) = k_{31}u'(x, y_1) + k_{32}\, u(x, y_1)$$

$$\beta_4(x) = k_{41}u'(x, y_2) + k_{42}\, u(x, y_2)$$

A special case might assign e.g.

$$u(x_1, y) = a_1$$

$$u(x_2, y) = a_2$$

$$u(x, y_1) = b_1$$

$$u(x, y_2) = b_2$$

where a_1, a_2 are constants or possibly functions of y, and b_1, b_2 are constants or possibly functions of x. Of course, G satisfies the same conditions from which the arbitrary functions are determinable. Writing

$$G_0(x_1, y;\ \xi, \eta) = a_1$$

$$G_0(x_2, y;\ \xi, \eta) = a_2$$

$$G_0(x, y_1;\ \xi, \eta) = b_1$$

$$G_0(x, y_2;\ \xi, \eta) = b_2$$

leads then to an easily solved matrix equation for the arbitrary functions where the matrix of coefficients of α_1, α_2, β_1, β_2 acts on the column vector whose components are $\alpha_1(\xi)$, $\alpha_2(\xi)$, $\beta_1(\eta)$, $\beta_2(\eta)$. The right hand side becomes the column vector whose components are $a_1(y)$, $a_2(y)$, $b_1(x)$, $b_2(x)$ minus the column vector $\bar{\Gamma}$ whose components are

$$\Gamma_1 = (x_1-\xi)H(x_1-\xi)\delta(y-\eta) + (y-\eta)H(y-\eta)\delta(x_1-\xi)$$

$$\Gamma_2 = (x_2-\xi)H(x_2-\xi)\delta(y-\eta) + (y-\eta)H(y-\eta)\delta(x_2-\xi)$$

$$\Gamma_3 = (x-\xi)H(x-\xi)\delta(y_1-\eta) + (y_1-\eta)H(y_1-\eta)\delta(x-\xi)$$

$$\Gamma_4 = (x-\xi)H(x-\xi)\delta(y_2-\eta) + (y_2-\eta)H(y_2-\eta)\delta(x-\xi)$$

5. SEPARABLE SYSTEMS

In solving partial differential equations, generally the first method learned is the method of separation of variables. First the partial differential equation is transformed into the coordinate system that fits the geometry of the problem. The equation is separated into ordinary differential equations which are solved, and finally these solutions are combined into a unique solution fitting the boundary conditions. The expression of the boundary conditions in a reasonably simple way is facilitated by coordinate surfaces fitting the physical boundaries. A valuable source book in this connection is the book of P. Moon and D.E. Spencer (Field Theory Handbook), Springer-Verlag 1971). We now have additional flexibility in that we can solve the separated equations or the original equation by the decomposition method (Adomian, 1983) yielding a further fertile field for exploration.

6. THE PARTITIONING METHOD OF BUTKOVSKY

In this work (Green's Functions and Transfer functions Handbook, A.G. Butkovsky, Wiley, 1982), any particular Green's function problem is partitioned into separate disjoint groups labeled by a triple of integers (r, m, n). Here r represent dimension of the spatial domain of definition D of the function u in a given problem, m is the order of the highest derivative of u with respect to t in the basic equation, and n is the order of the highest derivative of u with respect to space variables in the basic equation.

As an example, this equation

$$a\,\frac{du(t)}{dt} + b\,u(t) = g(t)$$

$$u(0) = u_0, \quad t \geq 0$$

is classified as (0, 1, 0). The equation

$$\frac{d^2u(t)}{dt^2} = g(t)$$

$$u(0) = u_0, \frac{du}{dt}(0) = u_1, \quad t \geq 0$$

is listed as (0, 2, 0), etc.

Basic equations are taken in an operator form exactly in the same manner as the linear form of Adomian's generic equation, i.e., $\mathcal{L}u = g$ or since we are speaking of deterministic linear equations $Lu = g$. Butkovsky is concerned with boundary or initial value problems for linear deterministic differential equations for a time $t \geq t_0$ and in a specified open set D with a boundary ∂D in Euclidian space, where g is a given function, and u is subject to given initial or boundary conditions. The problem is then said to be in standard form and characterized by a Green's function calculated and catalogued with the triplet numbers for about 500 problems!

Since Adomian's decomposition method allows the solution of nonlinear problems as well in the general operator form $\mathcal{F}u = g$ whether one, two, or more dimensions, the Butkovsky results can be used in the decomposition to solve the nonlinear equation. However, it is even simpler to use Adomian's suggestion for choosing an easily invertible part of the linear operator, usually just the highest ordered term. This makes the computations much easier.

7. COMPUTATION OF THE A_n

Let N be a nonlinear operator such that N(u) is a nonlinear function f(u). Cases where we may have for example f(u, u', u") and similar expressions are discussed elsewhere. Since the decomposition of u is parametrized so $u = u(\lambda)$ we have $N(u) = f(u(\lambda))$ which we wrote as $\Sigma_{n=0}^{\infty} \lambda^n A_n$. We find that A_0 depends only on u_0, i.e., $A_0 = A_0(u_0)$. Similarly $A_1 = A_1(u_0, u_1)$, $A_2 = A_2(u_0, u_1, u_2)$, etc., where

$$A_n = (1/n!)(d^n/d\lambda^n)f(u(\lambda))|_{\lambda=0} \quad (1)$$

which gives us a systematic, if somewhat cumbersome, computation scheme. Thus, if we write $D = d/d\lambda = (du/d\lambda)(d/du)$ then

$$D^1 f = (df/du)(du/d\lambda)$$

$$D^2 f = (d^2f/du^2)(du/d\lambda)^2 + (df/du)(d^2u/d\lambda^2)$$

$$D^3 f = (d^3f/du^3)(du/d\lambda)^3 + 3(d^2f/du^2)(du/d\lambda)(d^2u/d\lambda^2) + \\ + (df/du)(d^3u/d\lambda^3)$$

etc.

which becomes unwieldy. We observe that $D^n f$ involves $(d^\nu f/du^\nu)$ multiplied by polynomials in $(d^\nu u/d\lambda^\nu)$ where $\nu = 1, 2, \ldots, n$. For $n \geq 0$

$$D^n f = \sum_{\nu=1}^{n} c(\nu, n)(d^\nu f/du^\nu)$$

where $c(\nu, n)$ is the νth coefficient. Since $A_n = 1/n!\ D^n f\big|_{\lambda=0}$, we write

$$A_n = \sum_{\nu=1}^{n} C(\nu, n)h_\nu(u_0) \tag{2}$$

by letting $(1/n!)c(\nu, n) = C(\nu, n)$ and $h_\nu(u_0) = (d^\nu/du^\nu)f(u(\lambda))\big|_{\lambda=0}$.
The A_n can be computed directly from (1) or they can be computed from (2) evaluating the $C(\nu, n)$ from a recurrence rule which works very well. However, inspection of the results yields a procedure for simply writing out the results in a symmetric way which is computationally very convenient.

For any n in A_n we compute $h_\nu(u_0)$ for $\nu = 1, 2, \ldots, n$ by differentiating $f(u)$ ν times with respect to u. Thus A_3 involves $h_1(u_0)$, $h_2(u_0)$, $h_3(u_0)$. Since $A_3 = C(1, 3)h_1 + C(2, 3)h_2 + C(3, 3)h_3$ we only need calculate $C(\nu, 3)$ for $\nu = 1, 2, 3$. Suppose we consider $\nu = 2$ (and of course $n = 3$). We consider the sum of ν, in this case 2, and integers $k_1 + k_2$ or $\Sigma_{i=1}^{\nu} k_i = n = 3$ and ask how many ways can we add two integers to equal 3, taking only possible combinations not permutations. We have then $k_1 = 1$, $k_2 = 2$. Now C(2, 3) will be given as $u_{k_1} u_{k_2}$ or $u_1 u_2$.

Thus we take the product of ν u's with subscripts on u adding to n. What about C(1, 3) i.e., the coefficient of h_1? This must mean only a single u with a subscript 3, or $C(1, 3) = u_3$.

For h_3, we need three subscripts adding to 3, i.e., $k_1 + k_2 + k_3 = 3$ therefore we have $u_1 u_1 u_1 = u_1^3$. Whenever a repetition occurs, as in this case, we divide by the factorial of the number of repetitions. Hence $C(3, 3) = (1/3!)u_1^3$. Consequently, $A_3 = u_3 h_1(u_0) + u_1 u_2 h_2(u_0) + (1/3!)u_1^3 h_3(u_0)$. In general we take $\Pi_{i=1}^{\nu} u_{k_i}$ or $u_{k_1} u_{k_2} \ldots u_{k_\nu}$ with $\Sigma_{i=1}^{n} k_i = n$. If j is the number of repeated subscripts, we divide by j!.

Rules for the A_n have been given in Adomian (1983) and papers by Adomian and Rach. The method given here appears in a recent paper by Rach (see the Bibliography).

8. THE QUESTION OF CONVERGENCE

First let us briefly discuss a plausibility argument for the nonlinear case. For the equation $\mathcal{F}u = g$ where $\mathcal{F}$ is a nonlinear stochastic operator, we write a formal solution as $u = \mathcal{F}^{-1}g$. Now remembering we are solving *real* world applications (physical systems), we decompose the solution $\mathcal{F}^{-1}g$ into components $\mathcal{F}_i^{-1}g$ to be determined with $\Sigma_{i=0}^{\infty} \mathcal{F}_i^{-1}g$. We are *not* expanding in a series hoping to sum it to the solution u. We are *decomposing* u into components to be found just as one might decompose a function into impulses to determine the Green's function. Thus $f(t) = \int f(\tau)\delta(t-\tau)d\tau$ and the response to an impulse is the Green's function. We then integrate to get the response for the actual input. A rigorous argument is quite difficult and requires a number of conditions just as expansion in a Fourier series requires conditions. Our conditions are natural and satisfied in physical applications. The function we are looking for in the deterministic case is smooth; it does not have jump discontinuities. (In the stochastic case we have an equivalent situation with probability one.) In nature, jump discontinuities are rare. We may have a rapid transition and it has been mathematically convenient to represent such a transition as a jump discontinuity. Similarly the mathematician will speak of a process (the Wiener process) which is nowhere differentiable then demand proofs which will hold for these (physically nonexistent) cases (e.g., white noise). Real inputs are generally bounded and real processes do not have δ function correlations. Our methods are intended for real problems and we will proceed, bypassing the mathematical fictions for which the method doesn't work. For the applications, it is not necessary to include stringent proofs.[3] Pascual Jordan in the preface to his book Schwerkraft und Weltall has said that solutions to problems should be sought creatively without engaging in nitpicking proofs. We will introduce the theoretical concepts and proceed to solve the problems. We will often see that the solution is verifiable in a number of ways. Let us not forget that the rigorous theorems and sophisticated mathematical terminology so far have not solved the problems with which we deal. Usually one simply changes problems to simpler and solvable problems.

The "solutions" to problems in the journals invariably linearize nonlinearities and either ignore fluctuations or limit them to either being small or of a very special nature. Professor Wilson J. Rugh of Johns Hopkins University has said, "... when confronted with a nonlinear systems engineering problem, the first approach usually is to linearize, i.e., to try to avoid the nonlinear apsects of the problem. It is indeed a happy circumstance when a solution can be obtained in this way. When it cannot, the tendency is to avoid the situation altogether, presumably in the hope that the problem will go away. Those engineers who forge ahead are often viewed as foolish or worse. Nonlinear systems engineering is regarded not just as a difficult and confusing endeavor; it is widely viewed as dangerous to those who think about it too long ..."

Young scientists should realize that many pitfalls can be avoided by creative looks at problems and avoiding an overly scrupulous following in the footsteps of authorities. The Picard method, for example, is well known to every mathematical student. After calculation of a few terms, it becomes very cumbersome and the integrals become too difficult. Numerically calculating more terms does not give a better answer. Our method easily calculates as many terms as desired. Similarly, Itô's brilliant work was based on the only process which had received a great deal of mathematical study - the Wiener process - and must interesting mathematics resulted to explore the difficulties resulting from the mathematical fictions employed. However, randomness in nature may well proceed in a manner ordained by God rather than Wiener.

Physicists, engineers, biologists, and economists need to solve their frontier problems and that is what we will do unencumbered by mathematics for its own sake which sometimes tends to freeze approaches to the standard approaches. The latter happens for many reasons. We believe that's the way to do it because others have done it that way. Those who are experts tend to think anyone who doesn't do it the way they have for the last twenty-five years doesn't understand it. The young mathematician who wants to get grants must learn to play the game, or satisfy himself with principles rather than grants.

Certainly, it is important to know that attempts to compute solutions will be successful. Mathematically this means the problem is well-set, e.g., in the sense of Hadamard - that an operator exist which uniquely and continuously takes elements in a suitable class of initial data into a class of solutions. Statement of the precise mathematical conditions unfortunately requires a complicated symbolism not familiar to nonmathematicians, so we will simply inquire into the meaning. Solutions should exist for reasonable input data and each solution should be unique so it can serve as a physical approximation, and depend continuously without jumps on the given conditions. Also, it is reasonable to say small changes in parameters of the model should cause no more than small changes in our solutions. However, these problems arise generally from neglecting the nonlinear (or stochastic) effects or approximating them to first order so the mathematical solution does not remain valid for a long time. Nonlinear equations can be very sensitive to small input changes. If one linearizes a strongly nonlinear equation in his model then precisely defines conditions under which a mathematical solution to the simplified equation is valid, the solution of the model retaining nonlinearity seems preferable even if one knows merely that $u(x, 0) = g(x)$ and not that $g(x)$ belongs to a Sobolev space of L_2 functions with generalized first derivatives also in L_2. After all g represents a physical quantity. We are dealing with physical problems and the physical system has a solution and the parameters are generally well-defined without discontinuities.

For nonlinear systems such as those considered in this book, it is not generally possible to decide whether a given problem consisting of the governing differential equations and the set of boundary conditions is always well-posed in the sense of Hadamard nor is it obvious that Hadamard's postulates for well-posed problems are adequate to include physically meaningful solutions only. Our solutions require only the actual boundary conditions. Finite difference methods may require more conditions than normally required in an analytic solution.

NOTES

1. In the first author's usual form this is $Fp = Lp - Np = g$ where the inhomogeneous term is zero in this case. If $g \neq 0$, the solution has an additional term $\frac{1}{3}\left[L_x^{-1} + L_y^{-1} + L_z^{-1}\right]g(x, y, z)$.
2. They do in this case; when they don't, we can decompose each of the L_x, L_y, L_z into an invertible operator and a term taken to the right side of the equation.
3. A rigorous convergence proof is nevertheless appearing in an appropriate journal.

BIBLIOGRAPHY

Adomian, G., Applications of Stochastic Systems Theory to Physics and Engineering, Academic Press, to appear.
Adomian, G., 'On Product Nonlinearities in Stochastic Differential Equations', Applied Mathematics and Computation 8 (1981), 79-82.
Adomian, G., 'On the Green's Function in Higher Order Stochastic Differential Equations', Journal of Mathematical Analysis and Applications, 88 (1982), 604-606.
Adomian, G., Stochastic Systems, Academic Press, 1983.
Arfken, G., Mathematical Methods for Physicists, Academic Press, 1970.
Birkhoff, G., and Rota, G.C., Ordinary Differential Equations, Blaisdell Co., 1969.
Butkov, E., Mathematical Physics, Addison-Wesley, 1968.
Courant-Hilbert, Methods of Mathematical Physics (trans.), New York Interscience Publishers, Vol. 1 & 2 1953 -.

Hsiang, W.T. and M.K. Kwong, 'On the Oscillation of Nonlinear Hyperbolic Partial Differential Equations', *Journal of Mathematical Analysis and Applications* 85 (1982) 31-45.

Morse and Feshbach, *Methods in Theoretical Physics*, McGraw-Hill, 1953.

Rach, R., 'A Convenient Computational Form for the Adomian Polynomials', *Journal of Mathematical Analyses and Applications* (1984), to appear.

INDEX

GPSR Compliance

The European Union's (EU) General Product Safety Regulation (GPSR) is a set of rules that requires consumer products to be safe and our obligations to ensure this.

If you have any concerns about our products, you can contact us on ProductSafety@springernature.com

In case Publisher is established outside the EU, the EU authorized representative is:

Springer Nature Customer Service Center GmbH
Europaplatz 3
69115 Heidelberg, Germany

Batch number: 10607222

Printed by Printforce, the Netherlands